P9-AGN-570

About Island Press

Since 1984, the nonprofit organization Island Press has been stimulating, shaping, and communicating ideas that are essential for solving environmental problems worldwide. With more than 1,000 titles in print and some 30 new releases each year, we are the nation's leading publisher on environmental issues. We identify innovative thinkers and emerging trends in the environmental field. We work with world-renowned experts and authors to develop cross-disciplinary solutions to environmental challenges.

Island Press designs and executes educational campaigns, in conjunction with our authors, to communicate their critical messages in print, in person, and online using the latest technologies, innovative programs, and the media. Our goal is to reach targeted audiences—scientists, policy makers, environmental advocates, urban planners, the media, and concerned citizens—with information that can be used to create the framework for long-term ecological health and human well-being.

Island Press gratefully acknowledges major support from The Bobolink Foundation, Caldera Foundation, The Curtis and Edith Munson Foundation, The Forrest C. and Frances H. Lattner Foundation, The JPB Foundation, The Kresge Foundation, The Summit Charitable Foundation, Inc., and many other generous organizations and individuals.

Bet the Farm

Bet the Farm

The Dollars and Sense of Growing Food in America

Beth Hoffman

ISLANDPRESS | Washington | Covelo

Library of Congress Control Number: 2021936682

All Island Press books are printed on environmentally responsible materials.

Manufactured in the United States of America
10 9 8 7 6 5 4 3 2 1

Keywords: agricultural economics; agriculture subsidies; beginning farmer; BIPOC farmers; cattle ranching; contract farming; co-op; cover cropping; family farm; Farm Bill; Farm Transition; food system; grass-fed beef; grass-finished beef; hog confinement; organics; regenerative agriculture; rural development; sustainable agriculture; USDA

Hey, Dad, I did it.
I wish you were here to read it.

Contents

Chapter 1

The Simple Life?

Truth be told, I thought I knew a lot about farming. While I'd never farmed, I'd spent more than twenty years covering food and agriculture as a reporter. Traveling from California to Italy, India to Michigan, I'd visited farmers all over the world and learned about their lives, their struggles and successes. The idea of running a farm of my own felt like a gift, the culmination of my life's work.

And so a few years back, this suburban girl from New Jersey, living in big-city San Francisco, moved to a farm in rural Iowa, learned in her mid-forties how to build a fence, and made a fool of herself asking stupid questions like "Do the cows have names?" And although much about the environmental aspects of farming were familiar from my reporting days, the daily cultural and economic surprises were startling; I felt as if I had been cast in an updated 2020 remake of the old fish-out-of-water television comedy *Green Acres*. I quickly realized that reporting on farming and doing it myself were two very different things.

My husband, John, grew up on said farm in south central Iowa raising corn, pigs, cattle, and soybeans. In fact, John is the fifth generation of Hogelands to farm this piece of land, with its rolling hills, pockets of forest, tall prairie grasses, and, until the spring of 2019, acres and acres of corn and soybeans. As he puts it, "I spent summers immersing myself in the creeks and ponds, and winters tramping over the frozen, snow-covered hills. Spring and fall were for hunting and gathering, finding what the woods and streams had produced in silent bounty. All of my fiber knows that farm, and it fills me up with its life."

In other words, the farm is in John's blood. His great-great-grandfather James Ship Hogeland first came west from Indiana in 1851 as a surveyor for the railroad. It is said he loved the beauty of southern Iowa so much that he returned and bought land after his job was done. Two generations later the land was still in the family, the farm run by Lola and Pete Hogeland. Pete was also a banker, who, family legend has it, excused his neighbors' loans during the Great Depression in the 1930s, causing himself financial ruin. Then there were John's grandparents, Lloyd and Ellen Hogeland, who farmed the land until John's parents, Dorothy Lynn (known to family as both Dot and Lynn) and Leroy, arrived in the 1960s, took out huge loans, and together weathered the farm crisis of the 1980s. Hogelands worked and struggled on the same land generation after generation, forming stronger bonds with every nook and cranny of the landscape as time passed.

I, on the other hand, spent much of my life living in East or West Coast cities, going to plays and concerts, and frequenting museums. And even though I grew up in the Garden State, I did not know anyone who tended a garden, let alone a farm. "Land" to me meant a place where people built a weekend house in the country, not a multigenerational home where deep, spiritual connections to the creeks and hills were forged. While John immersed himself in the ponds of his family's farm, I went to summer camp at the Bronx Zoo.

So my move to a midwestern farm in middle-of-nowhere Iowa had many in my life wondering if I would be able to hack it. And to be honest, I wondered too. But not for the reasons they thought.

~

The first time Iowa entered my consciousness was the day I met John. We were neighbors in a three-apartment row house in Berkeley, California, where I had moved for grad school. He had just returned from visiting family in Iowa with his two young sons, he told me after introducing himself, a trip they took every summer. Iowa? I asked. Yes, he stated proudly—Iowa. In fact, he informed me, he planned to move back there and take over his family's farm as soon as his children were grown.

Over the next six months, John went from being an acquaintance to a close friend to something much more. When we married, the dream of moving to Iowa was still a good ten years off, and so I said I'd think about it. In the meantime, we spent weeks every summer on the farm with the kids, rocking on the porch swing, watching the lightning bugs, and swimming in the pond at the crack of dawn. We helped build fences and chased down escaped cattle, cleaned out old sheds, and picked chanterelles in the forest. I learned to drive the tractor around in circles cutting hay and went fishing in the dimming light of the day.

All the while, I was cultivating my own connection to the land and began to dream my own dreams of farming it. There was a certain badass feeling I got from the simple act of putting on my red rubber boots and strutting out into the fields to work, a freedom in sweating more than I had ever sweat, and feeling cleansed by it. And at the end of the day, there was a deep satisfaction in looking out onto the horizon at the new fence or recently moved cows chomping on fresh grass and saying, "I did that." I began to know in my bones that I wanted to spend more of my time outside, where my senses were alive with sounds and sights and at least some of the work I did was physical.

But there were several huge roadblocks in the way of our moving to Iowa, aside from the inevitable culture shock I was to experience. The biggest problem was that someone was already running the farm: John's dad, Leroy, was officially the man in charge and had been for more than fifty years. Although Leroy was already seventy-eight years old when John and I married in 2010, he was a spry seventy-eight, and he had no intention of simply turning the operation over to us. He had his (justifiable) doubts about whether I would ever really be persuaded to quit my job as a university professor and move to Iowa from San Francisco. And during John's absence from the farm for thirty years, Leroy had crafted his own plans for how the land was to be managed to support him in his old age.

The corn and soybean cropland was already rented out to a guy who sharecropped with him (an agreement in which the landowner and farmer share the costs and profits). Leroy still took care of a herd of cattle himself, and although he was interested in selling them, he, like many farmers, had no real plans of retiring. It was far more likely he would meet his maker while sitting atop a tractor than lounging on the beach or playing golf. Even if he could be convinced that taking it easy was a better way to live out his golden years, I for one couldn't imagine a time when we would be able to make our own decisions about the farm without his input.

To add to the complexity of the situation, like many beginning farmers today, John and I were entering our fifties, and although John was strong and I was fit, it was clear we were no spring chickens. I also knew well enough from talking to farmers through the years that the physical work of farming was in many ways the easy part. The challenge actually comes from everything else required for success. Instead of a romantic life of growing tomatoes and raising happy cows, farm life is actually a job full of spreadsheets, receipts, and file folders. Making a living as a farmer in America requires a lot of business know-how, in addition to the computer time, phone calls, and networking.

That's because, although the stories we tell about farms fail to mention it, farming is first and foremost a business. Loving the land or working diligently day after day in the heat and the cold goes only so far. To make the land your home and farming it your career takes more than passion or even high yields. It takes money.

I had seen with my own eyes how financially difficult farming could be and had met more than one farm family on the brink of losing it all during my years as a journalist. In 2016, a young farmer teared up as he told me that his farm had lost $400,000 the previous year and he would likely lose his family's hundred-year-old dairy in Michigan. I listened in 2009 to small-scale farmers in India talk of their struggle to buy food because they could not repay what they owed to the local moneylender for their rice crop. In 2012, an organic tomato farmer in Iowa told me of how her crop was dusted with herbicide from a neighbor's farm, ruining her entire harvest and throwing her into an expensive and lengthy litigation process to try to regain a tiny fraction of her lost crop's value.

But honestly, I didn't really understand how financial problems exist not just for poor farmers in developing nations or for a smattering of American farmers once in a while, but for the vast majority of them every year. I did not realize the amount of debt farms carry today—the average farm in Nebraska owes $1.3 million[1]—nor did I consider closely the challenges of a seasonal cash flow or the high cost of land. Like many privileged Americans when thinking about the failure of any business, I chalked up foreclosures and bankruptcies to ineptitude and a lack of creativity. Yet in reality, going broke is just over the horizon for the majority of farms in the country.

Leaving the city and heading for the simple life of the

farm, it turned out, is not all that simple. Financially, it is really hard. Farming is one of the most risky and expensive businesses one could start, and, increasingly with climate change, it is a totally unpredictable one as well.

But, I learned, American agriculture has always been this way. It has always been dominated by commodities—wheat, corn, cotton—sold to city dwellers or for export. Farmers have always struggled with overproduction and have taken the brunt of the risk and pain of low prices and devastated fields. Yet from the moment White[2] people set foot on North American soil, we have told romantic stories about farming, stories about family farms and their virtues, about independent farmers, about how farming can help save the planet. Stories about how the next new technology will increase yields, allow us to feed everyone on Earth, and help farms grow bigger and more successful.

Two dominant story lines—ones I call the "agrarian tale" and the "bigger is better" narrative—have not only confused the public about the realities of farming; they have also in effect trapped farmers in a limited understanding of their purpose, of the importance of their own economic viability, and of their relationship with larger corporations and with one another. Ultimately these myths distort the truth about agriculture in this country, making it hard to create something new, to form a system that is reliable *and* resilient, environmentally sound *and* economically viable.

~

So, I did wonder if we could make it as farmers, but not because we would lack easy access to the ballet and the closest sushi was more than thirty miles away. The most pressing question was how we could transition the farm from one generation to the next. Would Leroy ever let us take over? What would he think of our new ideas about how the farm should be run?

For me, even scarier was the thought that Leroy might actually say yes. I sat up in the middle of the night wide-eyed. What exactly were we getting ourselves into? Would we literally need to bet the farm and all our savings in an attempt to create a more ecologically and financially sound business? And what would farming really be like once we got past the romanticism surrounding it? John and I had huge dreams about what the farm might be, but would we have the chutzpah to go for it?

This book is our story of the Hogeland family farm (a term I will dissect later) and what the life of a farmer is really like. But through the telling of our personal story, the much bigger—and ultimately more interesting—story of the economics of American agriculture is revealed, a story with a long history I did not know before researching this book, even though I had reported on agriculture for more than twenty years.

I hope that in reading this book, you will come to recognize, as I have, that farmers do what they do *not* because they are evil, brainwashed, or, conversely, living saints. Corn farmers are not trying to destroy the environment or sacrifice themselves for the bottom lines of multinational agribusinesses. And small-scale organic farmers should not need to work for free to feed their community. There are economic, psychological, and pragmatic reasons for their choices, as well as cultural and political influences. Farmers today are pressed from all sides, and to avoid making the same mistakes we have already repeated as a nation time and again, we have to start examining those pressures. And while much of the conversation about food has turned into another divisive political blame game, we forget that systems evolve and can change course when opportunity, human ingenuity, and honesty align.

With this book, I want to talk frankly about the financial realities of farming, because farmers can make a difference only if they can make a living. They need to sustain themselves in order to sustain our land and food. From my first years on the farm, I believe that might be possible. Not easy, but possible.

Chapter 2

Land Rich, Cash Poor

A FEW YEARS BEFORE THE BOYS graduated from high school, we started talking to Leroy about leasing us the farm. We had spent a lot of time in Iowa by that point, and the prospect of farming had begun to feel like more than just a fantasy. But it was unclear how exactly we could take over when Leroy was still in charge and needed income from the farm. He hadn't planned for retirement and didn't have much in the way of savings, so we couldn't just show up and farm the land without paying for it. But if we were able to work out a rental agreement with him to ensure him some income, that would mean we would need to pay rent before we even started farming and long before we would see income of any kind from the farm. How exactly would that work?

Money wasn't the only issue. Because I had a good job in San Francisco as a professor and had lived all my life near big cities, Leroy was justifiably skeptical that I would ever actually move to Iowa. While John was able to return for long stretches of time, how and when he would be there full-time also remained to be seen.

And for the past few years, a younger farmer had been renting out parts of the farm to grow corn and soybeans conventionally (a term that indicates a farmer uses chemicals and, often, genetically modified crops). The guy arrived with his machinery and did things just the way Leroy had, using the same kind of equipment and receiving the same kinds of government payments.

Our request was different, and honestly, our plans for the farm once we leased it were pretty ambiguous too. John and I had no interest in growing corn or soybeans, conventionally or otherwise, and this departure from what Leroy considered the backbone of the business made him wary. We talked about growing small grains such as oats or barley organically, and John thought that buying the cattle from his dad was a good idea too. But there was no business plan, no spreadsheets comparing figures or making the case that we in fact knew what we were getting into. For Leroy, turning over full control of the farm for some vague, citified ideas was not only confusing; it was stupid. What if we floundered around and then couldn't pay the rent in a year? Would he lose the farm because of our naivete?

Years earlier, when John and I first got serious about our relationship, he had asked me to record interviews with his parents to document the generations of Hogelands on the land. John had two older sisters—sisters he adored and with whom I grew close over the years. But I didn't know much about Leroy's upbringing or about how he had met John's mom, Lynn (or Dot, depending on how and when you met her). She was a transplant from Galesburg, Illinois, and she ran the home after they married. Later, when she passed away in 2014, long after our talks, she was remembered lovingly for her affection for her kids and grandkids and for her delicious pies.

I sat at the kitchen table with Leroy and later with Lynn, microphone in hand, asking about what farming was like over the years. At the time Leroy was in his mid-seventies, born in 1932 and raised on the farm. He had taken over from his dad after returning from his army service in Germany during the Korean War.

Leroy is a storyteller, his memory packed with names and dates, important events, and childhood memories. He could talk for hours, weaving tales in amazing detail about his life. The interviews I conducted would create a great historical record, but they were also part of John's sly plan for his future wife to get to know the family quickly. The ploy worked.

"We had a huge garden and also milked cows, selling some of the milk in town," Leroy told me of his early days on the farm. He recalled being sent to town with a neighbor to sell a can of cream. "Dad told me that I could spend some of the money, which was a rare treat. I decided to splurge and bought a hamburger and a malt." But then one hamburger turned into two hamburgers, then three, and another malt. "And when we got back in the neighbor's car to head home, I threw it all up."

It was somewhere toward the end of the conversation that Leroy flipped the tables and put me on the spot.

What exactly were John and I planning to do with the farm? "Well," I said hesitantly, "we are interested in doing things a bit differently. Like maybe we will grow grains for humans to eat, organically."

"Organic?" Leroy earnestly questioned. "Do you have any idea of how much work that is?" He rose from the table and ducked out the back door. Moments later he returned with a fragment of a hoe in hand, its head worn down to an inch or so deep.

"This is what it means to grow organically." Leroy shook the flimsy steel nearer to my face, emphasizing his words. "Organic means that you are out every damn day hoeing and pulling weeds by hand. You want to do that with all your time?"

I didn't know. Was I really willing to trade in my city life to weed every day? And while the old hoe may have been a bit of hyperbole on Leroy's part, he also had a point. By far the biggest logistical difference between growing organically and conventionally is labor. If you are not going to use chemicals to kill weeds, you have to do it some other way—often by hand.

Leroy and other older farmers in the area I later interviewed all remember the days before commercially available chemicals as the Age of Weeds: the days before companies such as Bayer Crop Science and Dow Chemical Company became ubiquitous, when a hoe was one of the only tools in a farmer's tool belt for battling botanical menaces. They were days of much lower yields, when, if you were not diligent, weeds would take over the cornfields and garden beds. When herbicides arrived, many of them reported, the results were blissful—a horizon full of only the crop you planted. If weeds showed up, you sprayed instead of toiling for hours. To these producers, organic farming means a return to weeds, and pissed-off neighbors who see your weed-filled fields as a threat to their organized rows of corn and soybeans.

So, telling Leroy we intended to grow things organically was equivalent to waving a huge red flag in his face. I am sure he asked himself, "What does she know about growing anything, organic or otherwise?" Even though John had grown up on the farm, it now seemed he had been brainwashed by hippies in San Francisco.

~

Negotiations about us taking over the farm then took a lot of time. Years, in fact. Leroy would listen to John talk about his intention to move back and start farming, and he'd reply with a vague Iowa-style statement like "Well, John, I just don't know how that is going to work."

But the clock was ticking, and John needed to gain some ground in the discussions before I chalked the whole thing up to a nice dream and started making different plans. The boys were almost grown, and John's mom had passed away, spurring regret that he had not gotten back sooner to spend more time with her.

Leroy, on the other hand, had remarried and even at eighty-six still acted as if he might live—and manage the farm—forever, his sharecropping setup suiting him perfectly well. He was undoubtedly stubborn, a trait his son had inherited, much to my dismay, but Leroy would at least talk with us about our ideas and seem to listen to what we had to say. Signs that perhaps, slowly, he was growing more comfortable with our move.

Money, while not the only issue, was of course on everyone's mind, and once we were talking more seriously about the move, Leroy brought up almost immediately the idea of our buying his cattle. He would sell us the herd of twenty-eight mama cows and two bulls for $30,000—$10,000 a year for three years—lower than the market price but still a substantial amount of money.

But that begged the issue further: we would need land on which to graze the cattle. John would ask how much his dad needed for rent, trying to figure out a starting point for an actual lease, and Leroy would hem and haw. Finally Leroy produced pieces of notebook paper with calculations of the income he made from sharecropping and running cattle on the land, about $40,000 before taxes, a paltry amount considering that the debt he had carried for thirty years was paid off by then and the land was owned free and clear. But $40,000 was also more than we could ever pay in rent, particularly if we were going to go organic (a three-year process) or grass-finish our cattle (which would mean two years until we could sell a cow).

Yet we persisted. And slowly the discussions about leasing the pasture led to talks about using equipment. John found a lease template online and sent it to Leroy. Leroy and his new wife talked it over and made some changes, and the document came back our way. We signed and gave it back. After months, years, it looked as if an agreement were around the corner.

John flew back to do the final signing of the document with his father in person. Then, one afternoon when John was tinkering with a broken tractor, Leroy walked out back and announced, "John, I need to talk to you."

Leroy often "needs to talk," but the request always has an ominous feel. "What exactly does he want to talk about?" you wonder, bracing for the worst. Sometimes he just wants to make sure John knows something about the farm, like how the family fed the cattle during the 1934 drought or his ideas for growing soup beans. But at other times the talk is as foreboding as it sounds. This time he wanted to talk to John and both of his sisters. Together. Today.

That afternoon, the four of them gathered around the kitchen table in Leroy's house. "Take a look at this," Leroy said once they were all seated. He handed John a pad of paper. On it was a list of equipment. A 120-horsepower tractor with front-wheel assist and a rake. A hay baler. A bale hauler. A cattle trailer and a pickup to haul it. And more. All told, the list would have added up to more than a quarter of a million dollars if we purchased the machines used, and at a hefty discount. New, the equipment could easily have run us $600,000. The machinery would no doubt come in handy on the farm, but the purchase would break the bank before we even began our farming adventure.

"I've decided you need to buy all of this before March first," added Leroy, "or the lease will be null and void." A long, pregnant pause ensued. "What do you think about that, John?"

John looked at the list. Why was his dad throwing a wrench into the plans now, when we were so close to an agreement?

"Not much," John replied. One of his sisters took the list and read it too. "There is no way we can afford to buy all this," John added. The tension in the room was thick. The reasonable father who wanted to give his son a chance had left the kitchen; only a stubborn old farmer, stuck in his ways, remained. There was no negotiating with this Leroy; he had made up his mind.

"Well, it's my farm, and this is the way it's going to be," Leroy continued, slamming his fist on the table.

No one who was present remembers exactly what happened next, but more fists slammed and a few obscenities flew before John's sisters stormed out in disgust. "You don't want me to farm; you just want to tell me what to do," John recalls saying as he walked out the back door following his older siblings.

~

Our farm is not alone in spurring anguished debates; transitioning a farm from one generation to the next can be a heart-wrenching task. First, the older farmer—who has controlled the farm for the past thirty, forty, or, in Leroy's case, fifty years—has to be willing to allow the younger to take over. Giving up that kind of control is scary; many older farmers are not the retirement type. They are used to being constantly on the go; there is always a fence to fix or a tractor that needs a new part.

Siblings too often have different visions for what should

happen next to the land. One might want to farm, while others may want to lease or sell the land to the highest bidder, to take the cash and be done with it. I heard of a sixty-five-year-old son who had worked the farm with his dad for forty-five years. After his parents passed away, his siblings decided to sell the farm for a price he could not afford, putting their brother out of his lifelong career. Luckily in our case, John's sisters—Andrea and Alicia—both wanted to keep the land in the family and to have their brother farm it.

But what do you do if the farmer is still living, as in our case? Wealth on a farm is typically not found in the form of cash tucked neatly away under the mattress or in a 401(k) waiting for a farmer to turn sixty-five and a half. Wealth on a farm is locked away in its assets; farmers are usually "land rich, cash poor," especially now that land prices are so high.

In 2019, an average acre of farmland in California was worth $12,830.[1] Even in Iowa, the flatland of the former prairies in the northwest region of the state (land better for growing corn) can go for $9,300 an acre.[2] In south central Iowa, where the Hogeland farm is located—on land Leroy jokingly calls the "infertile crescent" for its erodible topsoil and sloped hillsides—a 530-acre farm is likely worth $2 million, even though the net profit one could make farming the same land would not come close to paying off the monthly mortgage.

Yet farms today have more than just the cost of land sucking up potential profits. In order to stay in the game, farmers pour huge sums of money into their operations.

Take, for example, the farmers who supply rotisserie chicken for Costco's new slaughterhouse in Nebraska. For that privilege, they had to spend $1.5 million[3] to build a facility made to the company's specifications—in which the farmers raise the company-owned birds as they are told. Personally, I couldn't stomach the anxiety of knowing my family would be indebted for the next fifteen years to pay off a million-dollar plus chicken house. And what happens *after* the building is free and clear of debt? If chicken farming has changed so much in the past twenty years, what kind of facilities will be required to raise chickens in the year 2040?

Even on a farm without contract farming, the price tag for basic equipment is mind-boggling. A *used* John Deere combine (albeit a top-of-the-line ride equipped with heated leather seats, yield and moisture sensors, and GPS tracking), commonly used in 2019 for harvesting corn or soybeans, can now easily run an operation close to $500,000.[4] For me, the cost of a $15,000 used pickup feels gigantic, but a $500,000 combine? I can't even imagine.

Yes, starting any kind of business requires a significant investment. To open a restaurant, for example, you need capital to build out a kitchen, decorate, and pay staff before you make a dime. But with a restaurant, a tech company, or even an online clothing shop, entrepreneurs look to differentiate themselves, to stick out and provide something unique that customers can't find anywhere else.

Farming corn or cattle, almonds or avocados for the mainstream, conventional market is not like other businesses in the United States. Farmers borrow money in order to grow the very same thing everyone else is growing, with the hope that being efficient and yielding as much product as possible will allow them to get ahead. It's a high-volume strategy, much like Taco Bell or Walmart uses—the margin so small for each individual chalupa or pair of socks that the company had better sell a ton of them. But even Taco Bell attempts to diversify by offering ten different beef-and-tortilla combinations, unlike most farms, which put all their eggs in one basket (pun intended) and then need to sell as many eggs as possible even though the market for eggs waxes and wanes.

Adding to the problem is that the more productive the seeds and sprays, combines, and confinement facilities become, the less profit you can make on each item. The technology ensures an ever-increasing supply of corn or wheat or soybeans, which, by definition, means lower prices when the supply inevitably outstrips the demand. Much like superpowers in a cold war arms race, the more farmers invest and produce, the more their neighbors also need to invest and produce, driving the price for goods lower and lower as more and more product floods the market.

Willard Cochrane, an agricultural economist who worked with President John F. Kennedy, called this the "agricultural treadmill," a debt-dependent spiral that ensures farmers continue to produce the same products because they are in too deeply to get out. It's a concept that Lynn, John's mom, discussed in my interviews with her in 2014:

> It is like a treadmill. You pay what you owe and you borrow some more, pay what you owe and borrow some more. Well, a lot of times we paid off the debt and there wasn't any left over or maybe still some debt. And that is the way it snowballed, just getting more and more, bigger and bigger. It is like playing poker—you hope what you make pays off the debt.

Farms need to stick with the new technology, even if it means little profit, because what would you do with a $1.5 million confinement facility but grow the chickens or pigs it was built to house?

The treadmill also works to keep farmers in the system much as a slot machine allows us to believe that, even though we have lost the past ten times we played, maybe next time we will win big. When prices are high, producing more seems logical, a way to get while the getting is good. Then, when commodity prices inevitably fall, you'd better produce more if you want to earn the same income. Buying an expensive combine can seem like just the thing to increase your yields, a strategy often promoted by extension agents and tractor dealers.

Farmers who come late to the party—who decide to put up that hog confinement facility after many others already have—struggle to stay afloat, Cochrane pointed out, the prices having already dropped as the market is flooded with product. "Average farmers . . . adopt the new technology, not to get ahead, but just to get back to where they were before. . . . Mr. Laggard, the last to look at the new technology, would be 'cannibalized' by his more progressive neighbors, an image strangely inconsistent with that of the hallowed family farmer."[5]

Leroy was no Mr. Laggard. It was the mid-1970s when Leroy and two of his neighbors decided to put up a silo, that quintessential symbol of the American farm. "Chopped" hay and corn was the thing back then, as opposed to big round bales, which are preferred today to feed to livestock, and a silo would mean that the trio could store hay as long as they wanted and feed more cattle to sell when their weight was optimum and prices were highest. Leroy took out a $14,000 loan (equivalent to about $80,000 today) from the local bank and built the silo and a small lot beside it in which to feed the cattle.

The building worked out well. So well, in fact, that in the early 1980s he decided to buy thirty-five calves to add to the forty head of cattle he already had. He went again to the local community bank and borrowed more money. But by this time many other farmers had also had the bright idea to put up a silo, likewise increasing the number of cows they were raising. And by the time Leroy's calves were grown, there was a glut of cattle on the market. The price of beef plummeted. In an attempt to pay off his debts, Leroy sold all seventy-five cows; the amount he made from the sale didn't even cover the loan.

Today the concrete lot still sits on the hillside amid rusting equipment and bent roofing, trees growing through it, all like a scene from *Planet of the Apes*. Right next to it stands the empty silo, the idea of storing corn and hay aborted after the cattle fiasco, the wind howling through it eerily, the tall concrete now home to birds.

~

We knew this history when Leroy demanded that we buy $600,000 worth of equipment. The game of farming is—according to Leroy and most of the neighbors—a gamble of trying to get out ahead and stay there. You borrow as often as you need to, and in farming you always need to. Even if Leroy almost lost it all because of his debts, for him debt is a necessary evil in farming.

John and I, however, had no interest in keeping up with the Joneses, in part because it is a system with questionable environmental consequences. But it also didn't make financial sense. Instead of investing in soon-to-be-obsolete buildings or in expensive machinery that eventually would end up with the rest of the discarded equipment—out in the field with trees growing through it—we wanted the money we put into the farm to stay there, to increase the value of the land and the *operation* as time passes. We wanted to find ways that the investments we make are not only agricultural but also ecological and social, investments to yield an increasing return for generations to come. The business itself should be worth something after all those years of hard work—the contacts and customers, the branding and the products—not just the tractors and a paid-off mortgage.

Instead of bleeding money up front, we wanted to invest strategically, finding the best bang for our buck. We saw ourselves as a lean start-up, a concept the tech world thinks it created but the rest of us just call making do with what you have. The idea is to *not* borrow money, to *not* become indebted to investors eager to gain a percentage of the business you worked so hard to create, and to instead figure out what it is you do best before you get too big for your own overalls. John and I knew we could make the fleet of old tractors work and could hire someone to bale hay for us for the time being. The decisions about pricey front-wheel-assist tractors would come later, when we understood what the land really needed and how we would pay for it.

That is, if we ever got the chance to farm.

Chapter 3

The Land of Corn and Cattle

JOHN, HIS SISTER ANDREA, AND LEROY entered the Beginning Farmer Center, a flat brick building near a Fireplace Superstore and a Chevrolet dealership in Ames, Iowa. The trio was there to meet with Dave Baker, director of the center and one of the nation's experts on transitioning farms from one generation to the next.

I had interviewed Baker as a journalist and hoped he might be able to help John and Leroy iron things out. "I think your dad just might not have any idea what we are talking about," I told John. Going organic or grass-feeding cattle must have sounded naive to Leroy, particularly coming from city slickers like us. Distraught after the "you need to buy a tractor" meeting with his dad, John finally called Baker, hoping that, since he was an official Iowa State ag guy, Leroy might actually listen to him. Leroy agreed, and they drove up to the Iowa State University

Extension and Outreach office with John's sister. Now all eyes were on Dave.

~

It's no wonder our ideas were completely foreign to Leroy—no one in the family, or in the county, for that matter, had done anything substantially different on the land since Hogelands first moved to Iowa in the late 1850s. Of course, there had been changes—Leroy's grandfather sent cattle by rail car to Chicago, a practice that has faded into history, and Leroy added a larger hog facility to the farm in the late 1970s. But over the 170 years that the Hogelands have owned land in the state, every generation has grown corn and fed it to livestock.

Learning that simple fact surprised me. When I arrived in Iowa for the first time some years earlier as John's girlfriend, I quickly realized that many of my unexamined notions about American agriculture had been completely wrong. For one, I recall being somewhat confused that there were birds chirping and that the frogs had only four legs; from what I had heard about the use of fertilizers and pesticides, not to mention genetically modified crops, it sounded as if nothing would be able to survive on an Iowa farm.

I also *thought* I knew that there was a "time before," a period prior to corporate control of the American food system when farmers grew a wide variety of products and sold them in town. There had been a time when everyone knew their farmers, I had imagined, and bought products directly from them, eating an all-local, mostly veggie-based diet. I had learned that it was only after World War II that large corporations came onto the scene and farmers started to grow corn, with the use of chemicals, machinery, and genetic engineering.

It turns out that much of that story is fiction. Although far more farmers used chemicals after the war, there never was a "time before" when all farms raised a diversified range of fruits and vegetables, livestock and dairy for sale at the marketplace. While families did grow produce for their own consumption, subsistence farming quickly became a thing of the past for American colonists; if you bought (or even squatted on) land in order to farm, you had to quickly specialize in a product to pay for it. Even in temperate places like California or the American South, produce could be harvested only at certain times of year and was hard to store, and transporting fruits and vegetables to faraway cities was risky in the days before refrigeration.

From the very beginning of White settlement, then, North American farmers raised commodities[1] such as corn, cotton, wheat, and meat to pay the bills, shipping their products to cities or to other countries. In fact, the first successful colony in 1607 in Jamestown, Virginia, was an enterprise funded by the Virginia Company to export products to England. They experimented with many commodities, finally establishing tobacco as the main cash crop by the end of the 1620s, with the first African slaves brought in to grow and harvest it in 1619.[2]

Corn too was grown in the coastal colonies. It was often used to fatten cattle, turning less valuable grain into higher-value protein in even the earliest days of White settlement. Maureen Ogle, in her book *In Meat We Trust*, explained: "Grain was in demand everywhere in the world, but . . . colonists learned early that the most efficient way to squeeze income from grain was by converting it into beef and pork. . . . Corn-fed cattle, in contrast [to grass-fed animals], arrived in better health and bearing more weight and returned greater profit."[3]

Boston, Massachusetts and New London, Connecticut became major livestock markets, aggregating animals from around the countryside by the early 1600s and exporting to even the farthest reaches of the British Empire in Asia.[4] Domestically too, colonial Americans quickly established a voracious appetite for meat. Ogle estimated that during the mid-1700s the average White North American male ate about two hundred pounds of meat a year (about four pounds a week) while their European counterparts ate such delicacies only once or twice a week.[5] Today, Americans average less than three pounds a week,[6] still a large amount for any nation.

Commodity agriculture continued to be the backbone of the American economy after the War of Independence, especially as colonists spread west. Compared with the limited space in Europe, the country's "open" tracts of land—although they were already occupied by Indigenous populations—made commercial production of food and fiber for export increasingly attractive.[7] Plows (high tech at the time) and machinery were used in the North and the slave trade expanded in the South as farmers increasingly specialized in growing commodity crops such as cotton, tobacco, and corn and purchased more of their food—including coffee, sugar, and tea—for use at home.

In 1848 the Chicago Board of Trade was created to formalize the buying and selling of futures contracts for commodities; its first contract that year was for corn. And by 1860 "globalization was already a major factor," wrote one historian; "the price of wheat . . . tied the American West firmly into the world economy."[8]

In the eastern midwestern states, including Ohio and Illinois, and a little later in Iowa, White immigrants swarmed in, and by 1860 the entire region had been divided into parcels and claimed by corporate entities and families alike. The delicate but deep-rooted prairie was plowed up to plant grain and potatoes, a plant used to further break up the soil. Corn was planted in abundance in the American Midwest as soon as White farmers moved onto the land. In 1860 the *New York Times* reported: "It appears that the great staple crop of the United States, not only in grain, but in everything, is Indian corn, with which no other country can produce a parallel. The total grain crop of 1849 (1850) was 867,000,000 bushels. . . . This is worth in market $450,000,000; double the value of all other grain crops, and more than double the value of the cotton crop."[9]

The homesteading land rush of the 1860s through the 1890s into Nebraska, Kansas, and the Dakotas was further fueled by dreams of striking it rich through commodity agriculture. The United States Congress granted land to railroad companies—often land that was not its to give, as defined by prior treaties with Indigenous tribes. These companies then sold the plots to aspiring farmers, frequently with false claims of fertile land and abundant water, pocketing the profits.

An example of this was in "bonanza farms," showcase lands that were among the first projects in the country to promote monocropping[10] (growing a single crop at a time) and used to lure prospective agriculturalists out west. The anomalies were written about in newspapers big and small; the *Atlantic Monthly*, for example, reported on these farms as a "revolution in the great economies of the farm,"[11] even though the majority of farmers would never have the kind of capital the bonanzas poured in to transform the land into a success. People eager to locate to the region were also encouraged to buy land near the rail lines so they could conveniently ship their crops by train, thereby again increasing the railroad barons' wealth as farmers paid top dollar both to buy land and to transport products.

Although some leaders, including Major John Wesley Powell, warned that much of the American Midwest and West—particularly the Great Basin—was far too arid to support large-scale farming, newspapers and Congress pushed not only for people to move but also for them to plow up the prairie to plant commodity crops such as corn and wheat and to raise livestock. "Come where you can get land. . . . Land that when you tickle it with the plow . . . laughs with its abundance," wrote one promoter, inspiring farmers to move to what was then Dakota Territory.[12] It was an invitation that not only ignored the inherently dry conditions of the area but in fact encouraged agricultural practices that made the landscape even drier, leading to the disastrous Dust Bowl of the early 1930s.

~

America's focus today on growing commodity crops is a direct result of our nation's history as a colony and, later, as a colonizing force when White agriculturalists moved west. The vast stretches of land made growing grains and livestock not only possible but essential—there was little other economic opportunity available to those homesteading on lands taken from tribal nations. Politically, in order to keep Indigenous populations from resettling, the land needed to be claimed and developed. An entire industry also sprang up to supply agriculture with increasingly large machines and innovative services, a newly patented device or tool arrived on the scene almost daily by the 1850s.

It is interesting to see this legacy if you compare, for instance, our rural towns in the Midwest and West with European villages. Europe was densely populated long ago, and although there was always trade (remember Marco Polo), space was scarce and travel difficult, especially in winter. There simply was not much room to grow a lot of raw materials or animal feed, and with areas often divided into fiefdoms, self-sufficiency in a smaller area was important. While in the United States railroad towns sprang up virtually overnight and were often gone as quickly, vibrant communities in Europe evolved over generations. Towns in Italy or Spain became known for a butcher who cut meat in a certain way, or a cheese maker who developed a particular variety made from the local cows' milk.

Specialty wines and fermented products developed in specific regions, often carrying the name of the town even to this day. Foods were consumed locally and were unique to each region.

Not so in the United States. Ottumwa, Iowa, is not known for a specific cheese, for example, as is Camembert, France. Nor is Lincoln, Nebraska, known for a cut of meat the way Parma, Italy, is synonymous with prosciutto. Instead of developing artisanal products, the expansive American landscape became an efficient supply chain in which farmers grew commodities, fed much of their crops to livestock, and shipped the cattle and hogs off to market. Towns evolved to be stops on the way from one point to another, the old town centers close to the railway (at the railroads' peak in the 1910s, Iowa was home to over ten thousand miles of track) or, later, along the interstate. "In North America," wrote Allan Nation, an expert on grass-finished cattle, "neither the farmer nor the consumer knew each other. It was a faceless relationship . . . separated by hundreds and perhaps thousands of miles."[13]

~

"My grandfather, Pete Hogeland, would ship a boxcar full of steers to Chicago every year or every other year at the least," Leroy told me when recounting the family history, a history that mirrors the academic research. Pete Hogeland likely began working the land in the 1870s and was a part of the corn-and-cattle food supply chain that evolved to keep city dwellers and Europeans well fed and midwesterners with money in their pockets. But that system had significant costs as newcomers en masse destroyed millions of acres of prairie and forest in order to feed the American agricultural beast. "Pete raised a lot of corn and fed it to steers," Leroy recounted. "But in order to get a lot of land under cultivation, he plowed up this land every year."

This is the flip side of America's efficiency and high production: corn plus cattle (and hogs) produced economic returns for speculators and large-scale farmers, but it destroyed the land in doing so. "This business of plowing up the land was disastrous for the country," continued Leroy. "A lot of this land out here was good land, before people came out and started plowing it up. But when they plowed it up, the ground washed away. My grandfather, Pete Hogeland, he made a lot of money," Leroy concluded, "but he did that by raping the land."

That language was telling. John and I proposing to grow something other than corn may have been heresy to Leroy. But despite his wariness about organics and grass-finished cattle, part of him knew it was time for change. He had seen the chemical prices creeping ever higher, the trade wars lasting longer, the weather becoming more severe and unpredictable. Fewer and fewer people would be left farming on the land, Leroy told me, if things kept going as they had for the past fifty years. "It's sped up now because the size of the machinery and the acreage that can be taken care of is so much bigger," Leroy explained. "It is going to end up with just a few people controlling all of agriculture."

"And what do you think about that?" I asked him.

"What do I think about that?" he pondered. "For the

average person in the United States, it is probably not too bad. Cheap food, maybe. But for the average farmer, it is disastrous."

Disaster is just what John and I would like to avoid, if at all possible. That's why we wanted to try a different kind of farming, one much less reliant upon expensive machines like the tractor Leroy asked us to buy.

~

And so there they were, John, Andrea, and Leroy, sitting in Dave Baker's office with a stalemate in our talks to lease the farm. Luckily for us, Baker had been around the block a few times with farm families and knew how to navigate the tough psychology of an older American farmer. Sure, Leroy would concede, things need to change. But given the state of agriculture today, what if we were to fail? What if the disaster so many American farmers have faced—namely, losing the land—befell our farm?

"I just think they are going to need a front-wheel-assist tractor," Leroy told Baker once the conversation was underway and John had explained how we'd almost had a lease signed, then suddenly didn't. The equipment he had listed was just the cost of doing business these days, Leroy argued.

But behind the discussion of hay balers and front-wheel assist, it was revealed, Leroy was really worried about what would happen if we couldn't make our cattle payment one year or didn't have the money for the lease. I had reassured him repeatedly, telling him that he would still own the farm and could rent it to someone else, but he knew that finding someone at the last minute to pay him the rent he needed might be a long shot.

"Leroy," Dave asked with empathy in his voice, "do you remember when you took over the farm? Do you remember how your dad wanted you to farm a certain way and you wanted to try new things?" Leroy nodded. "Did you make any mistakes?"

"Lots. All the time," Leroy responded.

"I did too," added Dave.

~

After Dave Baker worked his magic, things began to change. It was as if hearing from someone else that it was okay to let go was all Leroy needed, that the knowledge that other people experienced the same fears quieted them. Talks about the lease morphed into a back-and-forth with a new copy, the details worked out a little more each time. Finally, an agreed-upon document was on the table, and it was up to us to sign.

We signed in late February 2019. And then there was reality, ready to smack us in the face. We had been so fixated on the lease and convincing Leroy we really would move that we hadn't actually spent much time working out the specifics on our end.

Fortunately, the summer before, we had started converting an old red shed in back of Leroy's house into a very tiny house, about eighteen feet long by twelve feet wide. But there was still a lot to be done. The place had been cleared of eons' worth of family junk (John's report cards from elementary school, an old tuba, garbage bags full of old, ripped coveralls), and we had framed the space. The tiny house had insulation but no drywall, five new windows, and an old, rickety door that closed but let in bugs and cold air. In order to live there, we would have to work on it every day that summer and spend a fair amount of money on essentials, including a composting toilet, electricity, and a wood-burning stove.

While I stayed in San Francisco, John spent the month of March on the farm, seeding our new hay (instead of corn or soybeans) and attending to a slew of new baby calves. In a twist of terrible luck, our very first calf was born into a snow and ice storm and John found it stuck in the mud, shivering and barely alive. The calf died later that day, and we were both overcome with grief. Was this an omen? The universe telling us to stop kidding ourselves and give up before we even started?

But in early May 2019 we packed up the car with the necessities for another summer in Iowa—old jeans and light long sleeves, rubber boots, raincoats—and stuffed the old dog and her bed into the back. We rented out our house in San Francisco for three months to some high-paid tech interns and peeled off onto I-80. What to do about my job, the money we might not make, and our house, we would deal with later, when we had the time.

More urgently, we needed to focus all our attention on the farm. John named it Whippoorwill Creek Farm, a name that made our marketing-savvy friends cringe; it was not very memorable, they said, and too long. But we wanted to pay tribute to the creek that ran through it and to the bird of the same name we hoped someday would return. John wrote on our blog:

> I remember the call, can still mimic it with my own whistle, that one that I haven't heard in so long. As I grew up that call became less and less frequent until one summer, when I was eight, we heard only one lonely bird calling all summer for a mate, but no mate came. None have come since.
>
> So I chose the name Whippoorwill Creek Farm. It's long, difficult to fit into the small spaces paperwork allows, and requires frequent spelling over the phone and in person. But it is a beautiful name, a name that has a deep meaning and connection to this place I now steward. We are growing the trees and clearing the trash, eschewing the chemicals that make farming easier. I move my cows daily, not letting them trample too much or too long, fence them away from the streambed and banks of the creek, letting the wild riparian places return. For the wolves, bears, and mountain lions I can do nothing, though I would welcome them back. Against the raccoon and coyote the whippoorwill must make its own way; I can only supply places to hide.

John and I knew little about what we were getting into. But we did know we wanted to cast aside the long legacy of commodity agriculture to create something memorable and unique. We wanted to create a place where humans—not corn—could thrive, along with the whippoorwill.

Chapter 4

The Price of Sustainability

There was so much to do on the farm once we arrived in Iowa that our lack of planning faded into the background and survival mode kicked in. There were long stretches of permanent fence to fix, temporary fences to put up every day, and cattle to move daily. We had put in a fruit orchard with money gifted to us for our wedding six years earlier, and we had planted chestnut trees too, all of which badly needed our attention. There was the cleaning up of old metal and trash tossed into the creeks by generations of Hogelands in the days before landfills (or the money to pay for more proper dumping). And then, perhaps most important, there was the continued remodeling of the tiny house we'd started the year before so that we would have somewhere to live besides my sister-in-law's basement.

In other words, we were crazy busy those first few months. Busier than I had ever been as I also continued writing, setting up our blog, and engaging in a psychological battle with myself about whether I should quit my professor gig and move to the farm full-time. My mind was spinning, but whatever the specific worry of the moment, it was crowded by a steady drumbeat of anxiety—about how much money we were spending, how we would be able to afford health care, when we could move into the tiny house, and when and *if* we would start selling anything.

"Can you come pick me up? The wheel just fell off the tractor." It was John on the phone, his words matter-of-fact, as if he'd just chipped a tooth or gotten a big splinter. But I could hear in his voice that the wheel falling off was not as simple as it might sound. No, the subtext was more like "Hey, Honey, we have yet another very expensive problem on our hands."

The tractor in question was a big, red Case International Harvester. It was one of the only pieces of reliable equipment on the farm (given that we did not invest in our own, as Leroy had *suggested*), and at that moment, its main purpose was to cut hay. Haying is a life-consuming summer activity that has to be done at just the right time, in just the right weather, with just the right equipment, all of which can go off the rails at any point, costing your operation a boatload of potential income. If it is cut too soon, the nutrient content won't be good, and its value will be questionable. But if you manage to cut the hay at the right time and the weather moves in before it dries and you have time to bale it, the bales can end up wet. Wet bales can mold or, even worse, catch on fire and destroy more than just your bale.[1]

Like so much on a farm, growing the hay actually began months before, when John seeded the alfalfa, red clover, and a variety of grasses. This process happens every five-ish years to ensure the hay ground grows a rich combination of plants that people will spend good money to buy, if all goes well. It's an ecologically sound crop (although not as good as reseeding to prairie) that we can grow in Iowa without irrigation, chemical sprays, or fertilizer, a mix that grows like crazy on its own in the wet, not-too-hot summers and converts simple sunlight into nutrient-dense food for livestock.

Hay is also a cattle rancher's, goat herder's, or horse keeper's lifeline over the winter, or the summer in California, where, in an average year, there is no rain from March until October. All that hay, cut with the perfect nutrient content, dried, and kept dry over the rest of the summer and fall, feeds hungry animals. Often it is hauled to feedlots to be mixed with grains and other feeds (and antibiotics), or, as in our case, rolled out on the field and fed to our very own grass-finished animals when there is too much snow on the ground for them to find food.

In other words, the hay is worth a lot, whether as feed for our own cows—in which case we also get the benefit of keeping all of the nutrients on our farm—or as cash in hand when we sell the excess. So, a dead tractor in June meant that we wouldn't be able to cut hay at that perfect moment when the grass was ready and the weather was right. And that meant a potential huge loss of income that would have to be added to our already significant loss, due to the damn oats.

The hay-and-oat debacle started even before we officially took over the lease on March 1. We went in February to McCorkle's to talk about seed, that most basic of life-forms, which felt oh, so symbolic for us just starting out.

McCorkle's is an old-school farm store run by the family of the same name, open since 1919, where you can buy everything from gloves and coveralls to small buildings, fertilizer, and, yes, seed. The drive there led us down a series of frozen dirt roads lined with bare trees and brown fields, ending with a left turn into a collection of warehouses. To me the site is a modern miracle, a small family-run business on a dirt road in the middle of nowhere, able to survive and thrive despite the advent of Walmarts and Amazons.

First, we bought a single bolt in the hardware shop (my dad would have loved this place). Then we made our way over to meet with Kent McCorkle, the man in charge of seed sales. Our plan was to take the row-crop fields John's dad used for corn and soybeans and turn them into hayfields. This would allow us to become certified as organic, if we chose to go that route, since we would not be putting any chemicals on the land for the three years mandated by law. But it would still provide an income for us in the meantime.

We entered Kent's wood-paneled office, a space parceled off from the larger warehouse, and made our Iowa opening chitchat. "How's your dad?" Kent asked John. "How you liking Iowa in the winter?" he asked me. Then, once we were seated around his paper-filled desk, Kent started asking us the real questions:

What kind of hay did we want to grow?

Were we going to bale for horses? Dairy? Beef cows?

Big hay bales or small?

Were we going to broadcast or drill the seed?

Did we get the soil tested? How were the phosphorus and potassium levels? The pH?

What kind of soil? Clay? Sand?

And so on.

To give ourselves credit, we had some of the answers. We had tested the ground and knew the pH of our soils, and we also knew we were going to bale the large round kind, mostly for beef cows. Kent then explained to us many things we did not know: horses need more protein than cows, so you mix in more alfalfa; our phosphorus levels seemed a bit low; we could look online at maps made in the 1930s and 1940s to learn about the soil types on the farm.

John and Kent sat huddled together in the cold warehouse office in their lined Carhartt jackets, talking about how and when we were going to seed, the slope of the ground, the percentage of fescue we should mix into the hay. I sat on the side and listened, reflecting on how much my life had changed in such a short time and how much there was to learn about farming.

In the city, I often drooled over the seed catalogs that arrived in the mail, all the pretty flowers, the amazing cucumbers, the fifty varieties of tomatoes, pictures and descriptions enticing readers to buy way more seed than they could ever use in a home garden. But this was no wussy backyard garden we were talking about. We were buying seed for 110 acres—$8,000 worth of seed, so much seed that when we finally brought the first half home in the middle of the winter before we were living full-time in Iowa, it filled the entirety of our not-yet-completed two-hundred-square-foot tiny house. Yes, we forgot we needed to store the seed in a weathertight, watertight location where mice and birds wouldn't nibble away at our investment. At least the unfinished house served some purpose.

If there was so much to know just about seeds, how were we ever going to learn everything we needed to run the farm? It was literally our first moment as farmers, and I could see before us the thousands of decisions we would need to make, decisions that were all so complicated it made my brain feel like a jumbled mass of tangled electric fence.

Farmers have a certain knowledge that is hard for city dwellers to fully comprehend. It's not just that they intimately know a slice of the natural world less and less of us understand. It is also that the job description calls for lassoing very different skill sets, from financial, business planning, and engineering knowledge to an understanding of chemistry, biology, geology, and even climatology. To farm you must know how and when a seed germinates, what chemical components it needs, and where and how to plant it so that it will thrive. Plus, you need to be able to figure out the cost of planting and of running the machinery, and what, fingers crossed, you might be able to make back from it if all goes well.

John and I had virtually none of these skills as beginning farmers. John knew a bit of what his dad had done in the past, and we had some knowledge about the kinds of plants grass-finished cattle need in order to fatten on pasture without being fed any corn. So we lumped together our incomplete understanding of what we were doing with advice from Kent and other neighbors and did the best we could. We ordered up a mix of alfalfa, clover, fescue, and grasses and added another, less often planted seed to the mix—oats.

Oats are one of those save-the-world sustainable farming ideas you can read about everywhere, a cover crop farmers plant because the seeds sprout early and literally cover the soil, helping to prevent erosion in the late winter and early spring while the other seeds take their sweet time coming up. Today, cover cropping is a style of farming heavily pushed by virtually all parties, from the US Department of Agriculture to Unilever, from tiny organic farms to giant corporate ones, because when farmers leave land bare, erosion of fertilizer and topsoil is a massive problem.

And for organic growers, as we may be in a few short years, cover crops do more than just hold soil in place. They are a critical way to change up the crops grown on a piece of ground so that the soil doesn't become exhausted growing the same thing over and over. Organic farming by definition means you rotate crops in a multiple-year cycle, planting something different each year to add nutrients and complexity to the soil without adding chemicals.

But rotating crops or using them to stop erosion can quickly get expensive, especially when no one buys the mature plants. None of our neighbors use cover crops, as far as we can tell—the practice adds work and expense to a farm without a lot of clear economic benefits, many believe, although recent studies have found that cover crops do help financially.[2]

So, when the only organic grower in a hundred-mile radius told us he grows oats and *sells* them, we were thrilled. Before the other grasses and legumes grow tall enough, he told us, we can cut the oats and sell them as grain. We could earn a sizable bit more than we would with just the hay, making our bottom line for the farming season a lot less painful. Maybe with both the hay and the oats we wouldn't lose quite so much money in our first year.

It sounded too good to be true.

Years earlier I had been invited to a press day at a prominent energy bar producer's corporate headquarters. There, the owner of the company made an impassioned speech. Sitting in the company's small theater at the center of a facility that also boasted a farm-to-fork cafeteria, floor-to-ceiling glass windows, and a rock-climbing gym, the fit energy bar inventor talked about the company's challenges in making a 100 percent organic product. They could not source enough American-grown organic grains, such as oats, he told us, his voice strained with disappointment, even though his company would buy so much more if it were available. If we could just get the farmers to convert their fields to organic, he pleaded, the company, the farmers, and the Earth would all benefit.

Later that summer I attended a field day at Iowa State University. There, a group of farmers, businesspeople, and students stood in the pounding sun wearing wide-brimmed hats and sunglasses, squinting at the fields of corn before us. The researchers talked about their experiment. They were testing to see if growing organically would lower fertilizer runoff, a problem that has entangled the State of Iowa in huge legal battles[3] over pollution in the water supply of the city of Des Moines. They found that using cover crops, such as oats, as a rotation in organic farming created soil that could absorb more nitrogen, which meant less leached into the ground below.

The dots quickly connected in my brain . . . if farmers can grow cover crops like oats to improve the soil and the water *and* they can also make money by selling them to a company like the one I visited, isn't this a win for everyone? Shouldn't we all start growing oats, right away?

The researcher grimaced when I asked her why farmers don't just sell to the disappointed owner in California. "Those companies won't buy what we can grow in most of Iowa."

Other farmers and grain distributors at the field day chimed in. Companies talk about how farmers in the United States don't grow enough organic grain, but the oats we can cultivate are too "light" for them to purchase. The warmer temperatures in Iowa—particularly in the south—mean that grains don't grow fat enough to be considered prime oats, ones that could become your classic old-fashioned large, thick oats sold in grocery stores. The term "light" means a less mature interior seed surrounded by a lot more hull. Oats, it turns out, are not just oats, much as a steak is not the same quality as every other steak and cheese is not cheese. How things are grown, where, when . . . all impact the final product, even the lowly oat.

As with most things in life, behind the sad speech of the forlorn CEO there was a more complicated story, one that involves some pretty nonnegotiable things, such as the latitude of Iowa. It isn't necessarily that farmers don't *want* to grow organic oats; many of us can't grow the quality needed in order to sell them and make the whole thing economically viable.

Back in Kent McCorkle's office, John and I thought about the energy bars and the researchers. We thought of the neighbor who told us he grew oats not far away from our farm *and* sold them. Maybe we had misunderstood the grain distributors at the field day. Wouldn't our organic farmer neighbor know better? We hemmed and hawed a few moments and then bought the oats from Kent McCorkle.

~

About a month later John began the arduous seeding process, broadcasting the seed while the ground was still covered with snow. Miraculously, the oats did all they were supposed to do, ecologically speaking. They germinated earlier than their weed enemies, their little green shoots bursting through the soil, growing faster than the grasses. By May, at the end of the oat plant's short life, the fields had turned golden brown as the oats dried. The guy who was to combine them for us was excited; the spring had been cool and wet, he said, ideal for growing oats. Maybe the neighbor was right—perhaps we could grow great oats in southern Iowa.

But the cool and wet had a downside too. The oats were

not drying as fast as they needed to, and the grass, clover, and alfalfa were beginning to catch up, which would make harvesting the oats much more difficult.

Then the wind came. A storm was fast approaching, clouds stretching to the ground like a black veil traveling toward us from the not-so-far distance. We all stood on Andrea's porch (a very Iowa thing to do in a storm) as a wall of cold air enveloped us, the bushes in front of the house bending horizontally as if hit by a wave. Suddenly the top of a forty-foot-tall dead-standing tree snapped and crashed to the ground not far away. My sister-in-law, a normally cool and calm Iowan unfazed by extreme weather, hung her head out from the side of the porch to get a better view of the open fields to the north. "Okay, let's get into the basement," she stated matter-of-factly, the first time in all my summers on the farm that anyone had ever gotten nervous about the possibility of a tornado.

The five of us headed down; even the dogs who never came into the house were carried down the stairs. The rain hit the windows as if someone had turned a hose on full force, the sound against the siding as if someone had turned the volume up to eleven. Then, as quickly as the faucet had turned on, it switched off, and we all immediately went outside to see the damage. A few trees down, branches everywhere, but no tornado. No houses flattened. That night, the men did what men do and talked about how they never thought the storm was that big a deal, how they had all seen so much worse in their lives, how they couldn't believe we had actually gone into the basement like a parade of weenies.

But the oats had not fared so well. Many of the plants were now lying flat. Days passed as we waited for the ground to dry out again, the other grasses getting taller every hour. Suddenly we were faced with the question of whether we should bother to harvest them. The combine guy tested a patch a few days later, everyone's bubble now burst by the actual low yield of twenty-five bushels per acre, which barely filled the back of the grain truck. I climbed up the steep ladder into the bin and collected a baggie full of the oats, the smell strong like fresh bread, but greener.

The closest grain elevator was about ten miles away, a collection of metal silos and a red building with a truck scale outside for weighing the corn and soybean loads usually hauled to the facility by the semi. "Oats?" the woman behind the counter asked politely. She took the sample bag as she would with any farmer's grain and put it into the machine to test the water content and weight of the seed, the telltale signs of whether a grain would garner the highest price. "It's not really working," she apologized after a few minutes; "we don't really ever weigh oats." The owner was called. After a bit of fumbling around with the machine, a verdict was announced. The oats were still too wet and did not weigh enough for them to consider buying.

"We are having a harder and harder time selling oats from down here in southern Iowa," the owner told me. "We just can't compete with Canada and the northern states," he added, where oats seeds grow larger and heavier.

Was there another crop besides corn or soybeans that we could grow and sell in this part of the world? Would his company buy something else?

"Not really," he stated matter-of-factly. "You might think about growing rye, I guess. You could sell rye seed to other farmers for cover crops." I asked whether he had heard of Kernza, a perennial getting a lot of attention in places such as San Francisco because it produces a seed good for bread or cereal but doesn't need to be planted and harvested every year like corn.

He looked at me as if I were speaking a foreign language. "No," was all he said.

~

I learned later that we might have had a better chance of selling oats if we had bought a better variety and if we had seeded differently. If we had done more research about what kind of oats we could grow in southern Iowa and bought it from a specialty dealer who knew more, and paid more for it . . . it all might have worked out better. Or maybe not. Better, more expensive seed could not have changed the wind or the wetness of the spring, challenges that are part of farming no matter what precautions you take.

But we learned important lessons in growing the oats, even though at the time I might have preferred a profitable crop over more knowledge. We learned that seed variety matters and that what you can get at a typical feedstore in rural Iowa might not cut it for what we wanted to accomplish. We learned the necessity of understanding the existing market—who will buy your product?—and we learned that the way things are set up makes it difficult for farmers to grow a more diverse range of crops. And we learned to ask questions and then ask some more questions. The amount of things we didn't know, and didn't know we didn't know, was staggering.

The whole experience also made clear for me why most farmers buy genetically modified seeds and the sprays that go with them—it takes a lot of the guesswork out of the process and in the end makes a farmer's life much easier. Plus, the market for such products is readily available, and all you have to do is haul the corn or soybeans to the elevator and get paid. Growing anything else takes more time and money while also increasing your risk. It also requires far more extensive knowledge.

After weighing the words of the sad energy bar producer, the researchers, the organic farmer neighbor, and Kent McCorkle, after all that, we had made a choice that was good for the environment but sucked for us financially. And to add to the stress, the tractor was now broken. The tractor that was supposed to at least cut the remaining alfalfa, clover, and fescue and bale it into valuable rounds people would buy was out of commission.

I picked up John in the car on the side of the road, about a quarter mile from where he had left the tractor sitting cockeyed in the dirt. We sat silently as I drove us home, the financial weight of the broken axle heavy between us.

Chapter 5

Privilege to Farm

MONEY.

John was raised without much of it. "I got one pair of tennis shoes a year," he told me when I asked if he would say he had grown up poor. "We never went anywhere—going to Des Moines was a big deal—and we went out to eat probably twice a year." John could remember three family vacations as a kid: a drive to Colorado to visit distant relatives, another car trip to the Ozarks, where they stayed in a motel and picnicked, the third a trip to northeastern Iowa, "but I think that might have only been for the day," he added. The upside, he told me, was that they lived on a farm and always had beef, pork, and sweet corn to eat.

I, on the other hand, grew up solidly middle class in a three-bedroom split-level ranch-style house in a New York City suburb in New Jersey. My parents were art-loving people, and we regularly went to the theater and to museums in New York City. We ate out somewhat frequently, and we shopped at the mall when I wanted or needed new clothes. My parents paid for a fancy private college (Emory University) while John was at the University of Iowa on his own dime. We flew as a family to Florida to visit grandparents; John's grandparents lived next door to his family's double-wide trailer on the farm. My family certainly was not swimming in money when I was a kid, but we also didn't want for much of anything.

A few years after John and I got married, my parents helped us buy a house in a very modest and wonderful neighborhood of San Francisco for $700,000 (yes, that was modest in San Francisco), something we never could have done without help on the salary of a professor and a butcher in one of the most expensive cities in the world. We loved the house and our neighbors. And when we sold the place six years later, in August 2019, it went for more than twice what we paid for it. John and I had made a solid nest egg (based on my family's wealth) to take into our future.

Which didn't mean I wanted to spend everything we had on a farm in Iowa. As far as I was concerned, the farm business was its own entity and it needed to sustain itself without our bankrolling the operation as if it were some kind of expensive hobby. I was not up for being seen as the spoiled rich couple who moved into the area and tried crazy things like organic farming because, as the neighbors would likely say, "They don't need to make any money."

For his part, while John did not bring cash into our lives, the fact that his family had land was a game changer. Finding and accessing land is the number one issue for most aspiring farmers today—a cost so exorbitant that most people can't even consider farming unless they were born into a family already in the business. And while 55 percent of Iowa's land is rented out[1]—most often by families who don't want to farm but want to keep the land—it is hard to justify all the time and money needed to farm sustainably when it is someone else's property.

~

When you think about a family farm, you might conjure up images of young couples (often White) holding hands in a field, people cheerfully tending veggies, or those who own a few happy cows and a picturesque red barn. But family farms actually made up 98 percent of America's farms in 2020,[2] and they range in size from the smallest of farms to the largest. A family farm can be owned by retired siblings who live in the city and lease out the land, a large extended family that farms ten thousand acres of corn, or a young family with five kids that runs a hog confinement operation or allows their pigs to roam freely on pasture. Family farms can be of any size, with any type of production, owned by virtually anyone, making any amount of money.

And yet, to the American public, the term "family farm" is part of the romantic agrarian image. It means a farm that is wholesome, probably small, and maintained by salt-of-the-earth types who embody integrity and a strong work ethic. The term even seems to have a touch of conservative family values to it, as if farms keep a family together, even though few can afford to keep family members on the farm. Being fifth generation, like John, is a badge of honor that indicates your family has weathered some extremely hard times. But it also means that land has stayed in the same—mostly White—hands for generations, concentrating much of the nation's wealth. It is a system that is good for current landowners, but it means there is little opportunity for anyone else who wants to break into the business.

The Hogelands' ability to control this small piece of Earth is in many ways the result of decades of discrimination and displacement of others, as landownership in the United States has been closely tied to monetary wealth in the game of White privilege. The land we inhabit was once occupied by the Meskwaki (called Fox by the French and subsequently by the US government) and Sauk, formally taken in the New Purchase Treaty with the Sauk and Fox in 1842. The group was forced to live on the western side of the "painted or red rocks" (now Lake Red Rock, a popular recreational spot nearby) until they relocated to Kansas within three years.[3] In historian William Hagan's book on the tribes, he reported that tribal members felt so betrayed by the treaty that a group of them published a notice in an Iowa newspaper saying that their chief, Keokuk, did not act as their representative in the talks.[4] Leroy's grandfather told stories of Indigenous people still in the area when he arrived in 1851—three years after Iowa became a state and settlers poured in like honey to every nook and cranny of the region—but those tales are mostly of the sort I'd rather never repeat.

Yet after the two tribes—rolled into one by the US government—were officially moved to a reservation in central Kansas, some members stayed hidden in Iowa. Others soon returned. With the approval of the Iowa governor, in 1857 the group purchased its first eighty acres in Iowa (for twelve dollars an acre, not the ten cents they had sold it for) as the "Sac and Fox of Iowa," making the tribes private property owners. The federal government as a result stopped annuity payments to the tribe, claiming that it was illegal for American Indians to own land in the United States,[5] although these payments were reinstated ten years later.[6] Today the Meskwaki tribe owns more than eight thousand acres in Tama County in central Iowa and is the only federally recognized tribe in the state.[7] Yet although many Native people lived in the area around the farm immediately before the arrival of John's great-great-grandfather James Ship Hogeland, according to the agricultural census, today there is only one Native American farmer, whose tribe is not specified, in our entire county, hostility and discrimination having taken their toll.

Indigenous Americans are not the only ones who suffered a loss of land here in Monroe County, Iowa. About nine miles away from the farm once sat Buxton, a mining town established in the early 1880s. Monroe County had huge reserves of coal, an industry that once rivaled agriculture in the area, and company representatives traveled to Virginia to recruit workers for the mines. They brought hundreds of Black Americans to Iowa to fill jobs originally vacated by striking miners, much as they had done in other areas of the country.

Yet Buxton was unique. Once a mining town of five thousand people, the town was composed of both Black and White residents who, by all accounts, were paid the same wages, lived side by side in the same housing, went to the same schools, and were taught by both Black and White teachers. The town boasted Black doctors and pharmacists, lawyers, and owners of everything from farms to hotels. But surprisingly, by 1920, after miners were laid off and the company had moved on, virtually no Black-owned businesses remained. In Monroe County today, Black people make up only 0.8 percent of the population and live almost entirely in the town of Ottumwa. There is not a single Black farmer in the county today.

It is true that those who once lived in Buxton left largely because the company moved on. But people also did not stay because Iowa—like many places in the country—was not a very friendly place for Black people. In a PBS documentary on Buxton, Professor David Gradwohl recalled a statement by a former resident. "In Buxton we had our own lawyers, we had our own doctors and dentists and school teachers," the person had reported. "And then we moved to Des Moines and stepped back one hundred years."[8]

I visited the site of the old town of Buxton on a hot day in early August with Rachelle Chase, the author of two books on the town, and Jim Keegel, whose family now owns the site of Buxton. We stood atop the remains of the old company store, looking out at a few crumbling buildings peeking out of the weeds. Arguably, the field is just the site of another lost town in Iowa, one of thousands that dot the rural countryside. But as Chase described at the end of her book *Lost Buxton*,

> Buxton was started a mere 35 years after the end of slavery. Numerous African Americans interviewed stated that their parents or grandparents had been slaves, repeatedly sharing stories of their life of slavery. And those who had not been slaves still experienced extreme racism.
>
> They came from that to Buxton—a place where they could go anywhere they wanted, live any way they wanted, eat or shop where they wanted, and have the freedom they wanted. The freedom to live. The freedom to be. The freedom to choose. That must have felt like utopia.[9]

Yet that utopia was fleeting. Although Black residents owned businesses, homes, and acreage, that ownership didn't last, as it did for John's family. Across the country, Black families who defied the odds and acquired land (a.k.a. wealth) often lost it. The number of Black farmers in the United States peaked in 1920 at nearly one million,[10] but well-documented discrimination by banks and the US Department of Agriculture resulted in hundreds of thousands of Black farmers losing their land. The largest civil rights case in American history was settled in 1999 to compensate Black Americans for land and livelihoods lost as a result of discriminatory practices commonplace in USDA offices. Similar suits by Native Americans, Hispanics, and women have also resulted in hundreds of millions of dollars in damages.

So, even if Leroy got "lucky," as he described it, when he received a government loan in the 1980s to pay off farm debt (which, ironically, put him in a whole lot more debt), the truth is that being White and male certainly didn't hurt. John's family and others like them have been able to hold onto land in Iowa for decades, building family wealth along the way, not because they always made the right decisions or had better-run farms but because they received the loans and support necessary to put down deep roots. Those were advantages not afforded to Black and Indigenous farmers.

~

And so, while John and I had access to land and some cash to spend on our agricultural endeavor, we knew full well that many others did not. We were motivated to try to do something about the state of American agriculture, although, arguably, we were just an entitled Caucasian couple from San Francisco ignorantly trying to concoct solutions to society's ills. Yet it seemed to us we had an opportunity, maybe an obligation, to invite new people—particularly people of color—onto the land to share in our bounty.

We decided one way to change the status quo was to find people to farm alongside us in their own enterprises, an arrangement that would benefit not only beginning farmers looking to get a start but also the land and the overall productivity of the farm. A person raising poultry, for example, could move the animals into an area the cattle had just grazed, the chickens or ducks fattening up on the fly larvae in the cow poop while keeping the pest numbers down for the cattle. Bees could feed off the clover nectar and help pollinate the fruit trees, while sheep and goats could eat the invasive weeds in the pasture. Our farm would become an ecological nirvana with many different enterprises all working together but owned separately, creating jobs for many people along the way.

The vision was ridiculously idealistic. Tons of issues needed ironing out, first and foremost the fact that we had not really talked through these ideas with John's family. We would have to get their buy-in to have more people around the farm, farmers and their friends and family from all walks of life. Then we would have to find people, God knows where, who would be excited to head out into rural America to farm—and live—near a town that is 97.7 percent White and has basically nothing to offer except a gas station, a greasy spoon, and a dollar store.

There were also questions as to how exactly we would all work together—would we be a cooperative? Would we sell our items together or separately? There would need to be contracts and agreements, bylaws and lawyers. And to get a little morbid for a moment, what would happen to the group if we died in the middle of setting it all up?

But right now there were more urgent things to attend to than our fantasies of an ecologically and ethnically diverse farm, the end of racism in agriculture, and questions about the semantics of the term "family farm." There was the broken tractor, which luckily turned out to be one of the few things that went surprisingly smoothly. We found a guy to fix the axle, and within a few days he came out to the farm, saving us the trouble of figuring out how to transport the tractor to him. The repairs were not as expensive as we had feared, and in an amazing show of support, Leroy declared that, because it was actually his tractor, he would pay for half of the cost.

John literally rode off into the sunset on the newly fixed tractor. There was a lot of hay still to be cut—after which we hoped it would have time to dry completely before the rains came. There were cows to be moved the next morning and fence to be rolled out, calves to check on, and a pesky beaver to deal with who had clogged up the pond.

Maybe we would get time to think about the bigger picture sometime the following winter, when, hopefully, we could catch our breath.

Chapter 6

Money Matters

WITH OUR PIE-IN-THE-SKY IDEAS ABOUT how to change the face of American agriculture on hold for the moment, we had to get back to basics—how, exactly, would we make the farm economically viable? We had a rough sketch of a plan we had ironed out in San Francisco, put together on a spreadsheet with the help of my mother, Enid, a certified public accountant. But once we landed on the farm, we could see there were many factors we hadn't considered and costs we had to update in order to know if and when we might make a profit.

By May 2019, our bills were already sizable. The agreed-upon lease payment to Leroy of $13,000 had been made on March 1, and we purchased the oat and hay seed for the pasture over the winter, which ran us about $8,000. There was the $30,000 cost of the cattle we were paying to Leroy over three years ($10,000 a year), and of course there were endless incidentals—including the broken tractor—that kept our farm bank account steadily trickling out cash like a slowly leaking tire.

In late summer we finished our tiny house behind Leroy and his wife's home—a project that cost us about $10,000. It had no running water, but it had a great composting toilet (with no odor at all) and a five-gallon water jug with an electric pump that we refilled every other day or so for doing dishes. We cooked on a panini maker, had a device for heating water, and had a college-room refrigerator too. We installed an amazingly small but effective wood-burning stove—a cubic foot in size—to keep us warm and bought a two-seater sleeper sofa for our "living room."

We loved our tiny space, but it was, in fact, tiny. Soon after we moved in, I realized that spending months of a brown, muddy winter indoors in a space less than two hundred square feet while writing a book would probably be more than my psyche could bear. And although we got along well with John's sister Andrea and her family, driving the quarter mile up the road to take a shower at their house was getting old for everyone.

We decided the best option for maintaining our sanity was to buy a fixer-upper two miles down the road from the farm. The price was right—$40,000—but the hundred-year-old house had been used to house immigrant farm laborers, and sadly for both the people who had lived there and the old house, it was in total disrepair. The bathroom was moldy, with black growth between the tiles; the windows were covered in plastic in a futile attempt to keep out the cold. It took $120,000 and almost a year of work before we could actually move in, but in the end, the "salty dog" blue house looked awesome.

Aside from remodeling an elderly, broken-down house,

we had to figure out the logistics, and finances, of ranching ("ranching" is a term seldom used in the Midwest, but I use it here rather than "farming" to vary the terminology). The original plan had been to buy Leroy's twenty-eight cows and two bulls—a small number of animals for 530 acres but a good number for us to start with to figure out what the hell we were doing. Initially we thought that we would sell some calves at the sale barn nearby when the time came, but after taking one or two older cows there and seeing how it worked, I really didn't want to do it again. It wasn't that the sale barn was a bad place, but from what I could tell, the buyers were mostly from feedlots, and I wasn't interested in raising that kind of meat.

Going forward, then, we decided that all our cattle would be grass-finished, meaning that the calves would stay on the farm for a full two years, instead of the nine months Leroy kept them before sale, and would eat *only* what we grew on our pasture—no corn or other grains. (In using the term "grass-finished," I mean that the cattle are grazed *only* on pasture; some ranchers use "grass-fed" to mean the cattle have *access* to pasture but can also be given feed.)[1] John and I also wanted to refine the genetics of the herd, since the type of cattle that fattened fast on corn for the feedlot market, such as the Angus and Hereford crosses Leroy raised, were too big and tall to flourish only on grass. Plus, a few of Leroy's cows were downright mean—another trait we were anxious to lose in our herd.

To begin breeding smaller-framed cattle instead of our tall, leggy Angus-Hereford crosses, we decided to buy some Red Devons and Murray Greys. They foraged much better than the other cattle, eating a multitude of grasses and legumes, and were easier to finish on grass without grain. We located a guy who bred them, a Mennonite with the demeanor of a wise Zen master (who turned out to be a wonderful mentor), and bought twelve animals—a bull, two calves (one bull and one heifer), seven cows, and two yearling heifers[2]—costing us another $16,300.

But there remained the huge question of taste. People feed corn to cattle not only to make them put on weight faster but also because corn makes the fat marbled and the meat tasty. Grass-finished beef is leaner, and if the calves do not get enough protein, the meat can get tough and dry. Being a chef and a butcher in addition to a rancher, John was aiming to raise the tastiest cattle and took great care to be sure the heifers and steers[3] grazed on only the best pasture. But whether or not our meat would be up to our own high expectations remained to be seen. Aside from their shiny coats and rounded bellies, the cattle gave us few clues about their taste before it was all over.

Because of all that waiting around for something to sell, grass-finished beef costs more than the regular ol' supermarket variety—the price reflecting years of labor, fencing, winter hay, breeding, purchases of new cows, and the juggling of bulls. And how we would actually sell these animals once they were ready, whether wholesale or directly to consumers, we had no idea; that kind of marketing was going to be new terrain for us. Neither of us knew the laws regarding beef sales in the state of Iowa either, a hurdle we would need to jump sooner rather than later. In other words, money was steadily flowing out of the bank account, but not much business was replenishing the coffers;

we would be living off our savings for a while and hoping for the best.

We thought about establishing other income streams so we wouldn't have to rely entirely on the beef; selling only one product two years from now seemed like putting too many eggs in one distant basket. One idea was to sell the things that would grow for free in the forests around the farm, such as morel and chanterelle mushrooms, small walnut and swamp oak trees, wild plums, or tea made from bee balm and other foraged plants. There would be hay, of course, and John had planted about a hundred chestnut trees a few years back and a few fruit trees too, but they would not yield anything for years. We thought that eventually, someday, we would also have a place on the farm where people could hold events such as weddings and big, fun farm-to-table dinners. Maybe people would even stay overnight.

Generally, John and I agreed about almost all of the things we did and did not want to try on the farm, aside from the pawpaws he planted while I whined that it was a weird fruit no one had ever heard of, let alone would buy. We knew that we wanted to practice regenerative agriculture, a type of farming focused on growing deep-rooted perennial plants and grazing livestock on pasture for only short bursts of time in order to improve the soil. But how the rest of it might come together we weren't sure.

~

In farming, nearly every expense—whether you are growing corn or cows, tomatoes or strawberries—comes months, sometimes years, before any money flows back into your business. A farm orders corn or soybean seed around Thanksgiving and the chemicals that go with them in January, but the harvest of corn is not ready until October. Chicks must be selected and purchased along with a chicken house and chicken feed months before your first egg arrives. And if you want to go big, the scale of your operation can mean a lot more investment up front. Putting in strawberries at a commercial scale, for example, can cost up to $19,000 an acre,[4] the kind of money no farmer, regardless of the farm's size, has sitting around.

You might be saying to yourself, "So what?" All businesses have up-front costs. Opening a restaurant is an expensive proposition, as is starting a medical practice or a trucking business, all of which require sizable capital to get started. The problem is that, unlike these other businesses, after the first years of investment, farms continue to make very little money, often for generations. The median income for farms in 2019 in the United States was a whopping $296,[5] a positive number unlike those in the four previous years, as a result of $14.5 billion in Market Facilitation Program payments by the Trump administration to producers for the year.[6] And while farms that gross more than $350,000 often bring in more income than smaller farms, they require more capital to run, and often rely on astronomical subsidies to stay in business. The simple truth is that the vast majority of farms do not make enough money to support their own activities, let alone the families who run them—a problem I cover in detail in chapter 10, where I describe how farms actually do manage to stay afloat.

For now, suffice it to say that medium- and small-scale farms have a multitude of challenges to overcome (as do large farms, but their issues are of a different kind). First off is the huge problem of distribution. In most rural areas of the country—especially in the Midwest, where sizable cities are few and far between—there is no way for smaller farms to quickly or easily distribute their goods, no network of warehouses or middlemen set up to process, package, and move smaller amounts of goods into the hands of consumers. That means that we must figure out every detail of our own distribution along the way, including even the most basic logistics: Where will the meat be processed, and what kind of inspections legally need to be done? How will my product travel from the processor to the customer? How will it be kept cold? Who will drive, and in what vehicle? Unfortunately, volume doesn't always help either; the time it takes to figure out all the details often grows exponentially with each separate sale, as each order can be unique.

There is also the problem of seasonality. Everything has its season, as the Bible and the Byrds put it—a time to be born and a time to die—and although you would never know it from shopping at Kroger or Costco, pigs and chickens are actually not available year-round from the same region. So if Fareway or Kroger or Walmart were to decide to buy meat from smaller local producers instead of from Sara Lee (owned by Tyson) or Smithfield, as they do now, it would mean coordinating purchases with many different farms instead of with one giant packer, further complicating their supply chain.

Uniformity and consistency are also issues. Grocery stores might not end up with the same size of tenderloins if each farm raises a different breed in a different environment. Peaches grown without chemicals are often blemished or weirdly shaped; wheat processed in small batches and sold fresh might vary in color and be available only in the fall, none of which is acceptable to customers anymore. We want a tomato (or a banana) and we want it now, even if it is April in Iowa.

Being a small farm business also means you miss out on the economies of scale available to larger operations. A refrigerated truck, for example, costs the same amount whether you use it to haul one case of rib eyes or fifty, but your return on that investment is many times more if you pack the truck full and sell everything you carry. Even more profitable is hauling products continuously throughout the year instead of using the expensive truck for only three months when your tomatoes are ripe. And if a farm is big enough to grow not only tomatoes but also wheat and peppers, sales of tomatoes don't have to cover the whole cost of the truck—it can be deducted from the proceeds of many different products. The concept of economy of scale is also significant when it comes to marketing. Creating a logo and a website can be as costly for a small farm as it is for a bigger one; advertising to sell two hundred cucumbers may take a similar amount of time and cost as it does for two thousand.

But often the biggest hurdle for farmers, it turns out, is their own mindset. Struggle is largely accepted as a way of life, part of the mythology of farming; hard work should be accompanied with personal sacrifice on a farm, the agrarian story goes, trade-offs endured for the simplicity and peace of living on the land.

Back in the late 1800s, when peasant farmers did not get to choose their bucolic lifestyle as many of today's privileged farmers do, Karl Kautsky, a Czech-Austrian philosopher and hard-core Marxist, explained how it was that family farms continued to survive even when everything was stacked against them. "The progress of big industry does not necessarily entail the disappearance of small units. It ruins them, renders them superfluous from an economic point of view, but these units have enormous reserves of resistance," Kautsky explained. Families could sacrifice by eating fewer meals per day, for example, or by engaging in other forms of "self-exploitation" in which farms keep on producing food even when they do not make any money, "to maintain the occupancy of the land itself."[7]

No matter what you think of Marxism as an economic theory, it is easy to see how self-exploitation—often just to keep the land in the family—plays a major role in agriculture today, particularly on sustainable farms keen on improving the environment. Professor Ryan Galt of the University of California, Davis, cites Kautsky's quote in a report on the practices of more than fifty California farms practicing community-supported agriculture (CSA), a type of distribution system in which farms pre-sell products to customers (members) and deliver boxes of food throughout the season. The original idea of the CSA was that the community shared in the risk of farming by investing in the season ahead of time, sharing the bounty or lack thereof with the growers. What consumers got in their CSA box was a sampling of the harvest in real time; when cucumbers were ripening en masse, cucumbers would end up in the box, en masse. But today, Galt found, farms that consider sustainability or food justice[8] a top priority often don't pass on the true costs to consumers. Instead, some farms supplement boxes with food grown elsewhere, perpetuating a romantic image of farming by never revealing information about pests and disease, crop disasters, or labor issues. As one farmer put it, "The point of what we are trying to do is much bigger than to grow food and make money—I mean that's not even the point. It's to live sustainably and create communities that are growing their own food."[9]

This form of self-exploitation, as Kautsky and Galt called it, can be harmful to the farm in the long term, particularly if profits never materialize. "Values can drive you for only so long before the economic reality sets in that you can't afford the lifestyle you want to live," Galt explained to me. "Often it is posed in a binary way—either you are small scale and ecologically oriented, or big and a sellout. But I don't see the fetishization of the small as being a good thing economically for farmers."

Nor is it good for the industry. History shows us that new ideas, no matter how important for society, cannot survive unless they become socially and *economically* viable for participants.[10] If sustainable agriculture remains powered only by moral and ecological ideals—ideals accepted not only by consumers but also by farmers themselves—the change we are eager to see is not likely to stick. Galt argues, "Just because farmworkers and farmers—the work of both being fundamental to almost everyone's existence in industrial society—do not earn the average household income or more does not mean, ethically speaking, that they should not."[11]

~

You might notice that in the foregoing list of why farms struggle, laziness or a lack of wanting to work hard is absent. Quite the contrary; everyone seems to recognize that farmers, and farmworkers, work too long and too hard for too little. John and I, like so many other farmers out there, are far more likely to devalue our own labor and time by working way too much, just so that we can live a farming lifestyle, stay on the land, and pass it on to the next generation.

And yet hard labor is always given as a reason why sustainable farming isn't realistic, why more people shouldn't raise organic crops, and why there are so few younger people in farming today. "That's hard work!" A woman who has been involved in the pork industry for decades told me, citing her own millennial-age children, who, she said, would never want to do the grunt work required to raise hogs on pasture. Young people would rather have hog houses, she asserted, places where thousands of hogs are raised in a single building the size of a football field, than deal with the management headaches of raising animals outdoors and selling them directly to consumers.

I could be wrong, but raising thousands of animals in a building that smells terrible and produces millions of gallons of manure does not sound like a more enticing—or easy—job description than raising hogs on pasture. The real reason there are not more pigs rooting around in the forest, or vegetables grown in organic beds, or cheese made by hand, is that people don't want to work that hard *and not get paid for it.* John and I, and I believe millions of others, would prefer to live in rural areas, tending plants and animals. We are more than willing to work harder than we ever have before, even at fifty-plus years of age. The question for us is not whether we are willing to put in a full day's work but how we can afford to make improvements on the land, feed our neighbors interesting and tasty foods, and be paid a reasonable amount for our time and energy.

So I disagree with the farmers who say that money is not the point. In my opinion, sentiments such as Mike Madison's, expressed in his books about his farm in Northern California, are dangerous for the future of sustainable farming in this country.

> Those who set their prices high serve only the wealthiest stratum of the community. It also strikes me that a high price implies excessive self-esteem on the part of the vendor. . . .
>
> Because we both started out working as children, and for many years had low-paying jobs, we tend to put a low exchange value on our labor. . . . Our prices are low, and we get complaints from other farmers that we are undermining the market. And yet, we are unwilling to be farming only on behalf of rich people.[12]

Madison's idea—namely, that smaller farms should also be places that feed less affluent members of the community—is a tricky one. It sounds valiant to ignore costs and to say that helping people is more important than making money, although chalking a farmer's higher prices up to ego is, well, rude. But undercutting neighboring farmers who are trying to make a reasonable living—especially those farmers who have recently bought land and have not owned it for twenty or thirty years, as Madison has—seems shortsighted. Besides, who are the less affluent people in the community he is referring to? Farmers? Farmworkers? Restaurant workers? It is likely that the low cost of food is in fact contributing to the lack of money in the wider community. Even for poor residents who don't work in the food system, the answer is not cheap groceries (sold at the expense of the farmer) but a living wage that allows everyone to afford prices that reflect the real cost of production.

If we want new farmers to enter the profession, the economics of the farm need to be a primary, albeit not the only, metric for success. Farm families must be able to sustain themselves, to save money for college or to buy a house, to afford health insurance or a car. For many who want to farm, it is not enough to sink carbon in the ground or add nitrogen to the soil naturally. Leroy would not have been able to keep the land if he was not paid; the same is true for the next generation of farmers as well.

~

At the end of our first full summer on the farm, in 2019, John and I decided to sell our home in San Francisco. After much anguish, I quit my job at the University of San Francisco too, so that we could completely commit ourselves to our new farming adventure. And although John and I fell solidly into the category of the privileged once the money from our house was pocketed and we went to live full-time in Iowa, I wasn't interested in running the farm as so many others had, basically as nonprofits backed by family wealth and free labor.

The spreadsheets we concocted with my mother outlined a marginal profit for us of about $30,000 a year *after* about three years (if we didn't need to buy a tractor or another expensive piece of equipment), and since our new, old farmhouse would be paid for already and our kids were grown, it seemed like enough. Our cost of living was low in rural America, and farming would mean that we would have access to a lot of good food, if we did it right. Entertainment and eating out were basically nonissues, and there was no longer any reason or place for me to wear nice clothes and shoes. And with a little supplemental income—like most farms have (about which I will say more later)—we could be comfortable. Or at least we hoped.

Chapter 7

What about Subsidies?

SEVERAL YEARS BEFORE WE MOVED TO IOWA, John and I went to the county offices of the US Department of Agriculture's Farm Service Agency (FSA) and Natural Resources Conservation Service (NRCS). Located together in the tiny, struggling town of Albia, about a twenty-minute drive from the farm, the one-stop shop for farm programs sits in an unassuming one-story brick building as interesting as the flat parking lot that surrounds it. John and I walked into the cool of the air-conditioning and up to a long wraparound counter full of posters and pamphlets.

"Can I help you?" asked the woman at the counter, standing on the right-hand side of the office—the part occupied by the NRCS.

We described our situation. We were there to see if we could find any information about how to transition John's family farm to one with more diversity—maybe an organic operation with food-grade grains humans can eat, fruits and veggies, maybe grass-finished cattle. Could they help us figure out which government agencies might be able to help? Were there mentors? Pamphlets?

After a blank-faced pause, the woman spoke. "Your family's name?" She asked.

"Hogeland. John Hogeland."

"Oh!" Her face brightened. "Leroy's son."

Maps of the property were pulled out. Documents were clicked out of binders. An NRCS agent was called.

Yes, they could help. They could advise us on erosion prevention and tell us which of Leroy's lands were in set-aside conservation programs. They knew exactly how many acres of corn or soy had been grown on the farm for the past umpteen years, and—later, when we visited the FSA side of the building—they could tell us which payments Leroy was already signed up to receive. I am sure they also knew the expected average corn yield for the county, the soil types we had on the farm, and how many acres had been used for grazing in past years.

But where might we sell oats or barley if we grew them? Organics? They stared at us like the proverbial deer in the headlights. "There's a guy in the next county over who grew something organic," an agent actually told us. "Maybe you could ask him for some advice."

Not a single person in the office could talk to us about growing apples or cherries—fruits that actually grow well in Iowa and historically were grown in great numbers—nor could anyone advise what small grains might do well in southern Iowa. Nuts? Not a chance. There wasn't even anyone at the office who knew anything about pasture raising pigs or chickens in the county.

The same scene repeated itself over the years. Each time we arrived at the counter, the person behind it would recognize John. An FSA agent would bring a set of forms, already filled out, for him to sign and discuss deadlines for the next season. An NRCS employee would remind us how much time was left in Leroy's CRP contract (the Conservation Reserve Program[1]—a program designed to remove land from agricultural production for a period of at least ten to fifteen years) and how much money had been approved as a cost-share for us to put in new fencing that fall.

But every time we said the word "organic," a look of mild discomfort would come over the person helping us. "We don't have any programs for organic, per se," they would tell us. We would reply that we had heard there was. The woman would disappear and return some time later. "No, I asked everyone, and no one knows of anything available specifically for organics. It is just the same programs as the other farmers get. If you want to, we can talk to you about putting land into a conservation easement."

The truth is that programs do exist. There is a cost-share program for transitioning to organic, for example, that no one in the office knew about. And many different NRCS programs directly support the environmental upgrades necessary for organic certification, such as rotating cattle on pastures. Other programs promote better ecology more generally, from putting in bat habitat to establishing oak savannas and improving waterways. These programs directly affect ecology by giving farms the money to put sustainable practices into action.

To be fair, no one in the county had likely ever asked about growing nut trees or transitioning to organic, so why would an overworked employee spend her free time learning about those programs? Agents in Sonoma County, California, I am sure, know all there is to know about organic programs because so many growers there are certified as such, but I'll bet they would be hard-pressed to tell you much about the supports available for growing soybeans.

But it was 2018, after all, a point in our agricultural history when "organic" had become a household term, with organic products available even at Walmart. Growers in this part of the world might be interested in learning more about organics, or at least want information about growing small grains or doing rotational grazing. If the agents in our county knew more about these currently less popular programs, it could expose farmers to new and different ideas, to techniques they had not thought of, perhaps even to new crops that might in the end be more profitable for them.

For the employees working at the USDA, part of the issue is that the programs constantly change. Every five years a new Farm Bill is introduced,[2] the legislation covering everything from food assistance (such as food stamps, officially called the Supplemental Nutrition Assistance Program, or SNAP) to conservation and subsidy programs. After the grueling political debates end and the legislation is finally passed, the job falls to the USDA to interpret the law and figure out how to get resources to farmers. That part of the process can take a year or more, at which point programs start popping up at the local USDA offices along with new guidelines. A chunk of money must be given out by a specific date, offices are told, encouraging agents to find people quickly to receive the money, and new applications are processed—often according to detailed ranking systems staff must also learn how to navigate. In 2019, there was also a new computer program for employees to use, which presented its own challenges. It's a top-down system that leaves employees little time to work with farmers one-on-one to find solutions to their problems instead of simply taking applications and entering data into a computer.

~

I returned to the USDA office again in 2020 after we had fully moved to the farm, two years after the 2018 Farm Bill had been passed and the office was just getting caught up on how to implement the changes in it mandated by Congress. I was there to take a class about the complicated and confusing programs we were eligible to sign up for, an hour-long presentation to be given by our local FSA head.

The programs in question were Price Loss Coverage (PLC) and Agriculture Risk Coverage (ARC), two of the main supports available for corn and soybean farmers in our county. The first step, I learned, was for growers to discern whether their farm might be eligible to receive a payment at all. Payments are not given for *planting* commodities; they are given for crops historically planted on the land: a designation called "base acres." Of the Hogelands' almost 530 acres, for example, 150.6 are registered as corn base acres, and 25.7 are soybean. Even if we planted corn on every one of our 530 acres—or, conversely, on not a single acre—we would receive a payment for the 150.6 acres designated for corn. The same is true for soybeans. The FSA agent explained that farmers are paid regardless of whether the commodity is planted or not (in what is called a decoupled payment) so that they retain the "freedom" to respond to the market; they do not plant a crop just to get a payment. If the supply of corn in the world is too high and the price is low—the theory goes—farmers will not plant corn, even if they have corn base acres, because there will be little demand for it. Instead, they can plant soybeans while still collecting a payment for corn.

In the real world, things don't work this way. Corn and soybean rotations are in fact planned years in advance. NRCS agents provide maps of which crop should be grown when and where on a farm, and farmers who want to enroll in many of the programs must follow these recommendations. Farmers also typically do not plant corn in the same fields two years in a row, not only because the NRCS does not allow it but also because the crop will not do well; growing corn takes a lot of nitrogen, which soybeans help to replace.

~

Each of the two programs—ARC and PLC calculate payments in a different way. PLC payments kick in when a predetermined price for the commodity (again, only for base acres) is not met by the national average price for the year. ARC payments are determined by taking the *average* yield for one's county and subtracting the *actual* county average yield for the year. Corn and soybeans are, of course, not the only crops supported by these programs; twenty-two commodities are eligible, including chickpeas, peanuts, lentils, and sunflowers.[3] These subsidies, whether based upon the yield or the market price, mean that the payment a farm receives can vary greatly from one year to the next.

Yet the largest government support comes not from direct or indirect payments but from crop insurance, the second-largest expenditure of the Farm Bill after SNAP, which takes up 75 percent of the spending. In 2017, the $5.6 billion program paid more than 60 percent of the premiums for farms and another 20 percent of the premium costs to the insurance companies for administrative costs.[4] Again, the big winners of the insurance programs were the usual suspects: although more than one hundred crops qualified, more than two-thirds of the acreage enrolled in the program was planted in cotton, wheat, corn, or soy.[5]

To add to the regular old payments that commodity farms usually get, however, the Trump administration added $23 billion[6] in total payouts under the Market Facilitation Program (MFP) to offset the pain of booming and busting markets and "unjustified" trade wars, most notably with China. (According to the World Trade Organization, the United States was involved in 280 active trade disputes with at least ten different countries as of March 2, 2021.)[7] The MFP money came from a USDA account that did not need to be authorized by Congress, and in two years it paid out more than the amount spent to bail out the auto industry in 2008.[8] Yet as with the other payments, most of the money went to the very large farms. An analysis of the data found that 10 percent of farms in the United States received almost 60 percent of the payments, with many receiving well over $100,000. Eighty percent of recipients received less than $10,000.[9]

~

There is always the lingering question of why we don't instead subsidize table foods—the fruits and vegetables we should all be eating more of—instead of paying farmers to produce commodities such as corn and soybeans. Shouldn't we be paying farmers to grow food with fewer calories and more nutrients?

Back when subsidies were first introduced during the Great Depression, fruits and vegetables were not included because they were not mass marketed and therefore not impacted by the huge booms and busts of the international stage that had caused so much chaos for farmers. Much later, in the 1990s, when decoupled payments were first introduced, fruit and vegetable growers fought to exclude specialty crops—fruits, veggies, and a whole mess of other crops including herbs, potted plants, trees, and shrubs—from the subsidy structure. They feared that if fruits and vegetables were part of the same system as corn, wheat, and cotton, farmers could get paid for their corn base acres but choose instead to grow cherries and kale. They could sell the produce for less than it cost to grow because it would, in effect, be subsidized. So the law specifies that the decoupled payments *do not* apply if farmers

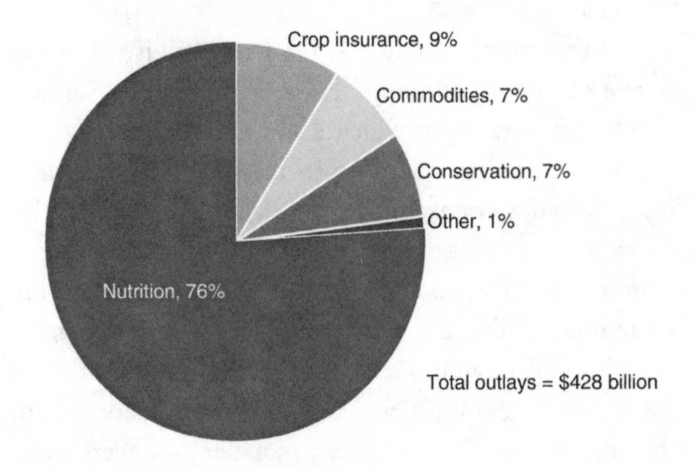

Sources: USDA, Economic Research Service calculations based on Congressional Budget Office estimates.

Figure 7.1

choose to grow specialty crops on their base acres—if they decide to grow blueberries or apricots, for example, they lose commodity payments for that acreage.

A host of grants are aimed at specialty crops to support things such as research, education, marketing, and food security, but funding for such programs is nowhere near the money spent on commodity supports.

Commodity payments and crop insurance (also targeted at commodities) total 16 percent of the total budget (see figure 7.1). That comes to almost $68.5 billion a year for crops such as corn, alfalfa, rice, and wheat. Specialty crop grants and other programs, including money for farmers markets, get just 1 percent of the total funding, or $4.28 billion a year.

It is important to note that the Farm Bill isn't the only source of funding for crops. Some are supported through checkoff programs, in which a few cents of every dollar spent on a product goes into its marketing and research. Hass avocados, for example, have a checkoff program that growers pay into, as do raspberries, mangoes, and honey;[10] famous advertising slogans, such as "Got Milk?" and "The Incredible, Edible Egg," came out of checkoff programs. But checkoff programs are controversial in that, although 100 percent organic growers with certification are exempt, the programs don't really cater to smaller producers trying to sell products locally. A small-scale avocado grower for example, would be charged the involuntary tax, but her farm arguably does not benefit from, and is potentially hurt by, money used for research on how to transport avocados over long distances. John and I were charged checkoff fees the few times we sold cattle at the sale barn, the money going to "stimulate restaurants and grocery stores to sell more beef and encourage consumers to buy more beef,"[11] although both of us feel strongly that people should eat less—albeit better-raised—meat.

There is also a practical problem with supporting farmers in growing large amounts of fruits and vegetables: they are perishable, which makes the products difficult to transport and store. If a farm or a nation ends up with an excess of corn, it can be put away for later use. Not so with broccoli. And in a national emergency, with crops failing because of a huge drought, for example, reserves of wheat and peanuts can be used to fill bellies more readily than we might be able to distribute zucchini. This perishability also means that for midwestern farmers with hundreds of acres of land, far from urban populations, there simply are not as many options for crops to grow. Even if the machinery they own could be used to grow greens and carrots instead of corn, to whom would a farmer in deep rural Iowa sell hundreds of acres of vegetables?

So, even though we did not plant corn or soybeans—and had no intention of ever doing so—Whippoorwill Creek Farm received $4,554 in commodity payments from the government in 2019.[12] When I asked at the FSA office what we should do if we didn't want the payment, the woman gave me a baffled look and shrugged. Ten years ago, when Leroy did the farming himself, the payments he received for growing commodities were similar, ranging from about $3,000 to $8,000 a year, depending on the markets and yields, according to publicly available data.[13]

These payments are typical for our county. While commodity payments fluctuate widely year to year, many landowners in this area of rolling hills and forests don't receive a whole lot in commodity payments, even though when you drive through, it looks as if there are acres and acres of corn. Instead, people here enroll heavily in conservation programs, mostly the Conservation Reserve Program, and receive a steady annual payment for *not* farming land. CRP payments in the county totaled $6 million in 2019, while in the same year farmers received about $2 million in commodity payments.[14]

But in the state, commodity payments get much, much bigger as you travel into flatter terrain, where corn and soybeans have historically been grown on more acreage. The largest payment in Iowa in 2018 went to a farm about a forty-five-minute drive from us, with the notable name Iowa Family Farms. According to USDA data, the farm received almost $800,000 in commodity payments and another $142,000 in CRP payments that year. Interestingly, two years before, in 2016, the owner and operator of the farm, Patrick Hammes, was awarded second prize in a national corn yield contest, hitting 320 bushels per acre on his best field.[15]

To me this is an important point: despite the myth that independent farmers believe the government should stay out of their business, even farms with the largest yields are not self-sufficient enough to function on their own without government assistance. And with the return on investment for larger farms at only 4 percent *with* government payments[16] . . . is there such a thing as a successful large-scale farm?

At this point in American history—almost one hundred years since the Great Depression and the beginning of agricultural subsidies—it feels safe to say that commodity farmers expect the government to pay them—and some of them very handsomely—if a new trade deal or a new law (mandating corn ethanol use, for example), the weather or the market doesn't work out. Farm payments have in fact become entitlement programs—programs in which a vulnerable group—farmers—receives protection and support regardless of poor business decisions made. One of our neighbors boasted that his favorite president was Bill Clinton, not because he helped make farming more profitable or more stable but because during the Clinton years, the farmer got two payments a year instead of one. Those were the good old days.

For the average 359-acre farm in Iowa,[17] these different supports don't provide the kind of income that would convince a farmer to grow commodities. But they certainly are not a deterrent either. The payments are large enough to be useful, helping to pay off loans on farm equipment or seed, but not of the type that will save a farm when times really get tough.

But these supports provide farmers with a lot more than money; they are culturally important. The USDA support acts as a sort of "we've got your back" endorsed by the American people. There are offices full of people to talk to about your crop, university professors working to be sure your planting regime is the best possible, and companies spending billions of dollars each year to create a seed for you that works. So the real question is, why would you not plant it?

The fact is that even if subsidies farmers disappeared tomorrow, farmers would still likely produce corn and soybeans in Iowa, although arguably not as much, simply because they don't currently have many other choices of what to do with the land. We are not a nation that consumes large amounts of barley or rye, for example, crops that could be integrated into the corn-soybean rotation to help improve the soil, reduce pests and weeds, and lessen the need for fertilizer. Nor is much of the Midwest populated enough to handle the production of large quantities of fruits, as I pointed out earlier. Corn and soybeans grow well in southern Iowa, while—as we found out—oats typically don't, and vegetables can be plagued with pests in the humid climate, making it tough for a farmer to compete with drier places such as California.

So what is a farmer to do?

~

By the end of the hour-long class at the FSA, I was both better informed and more confused than when the session began. I had learned that even though we weren't going to grow corn or soybeans, we would get a payment, if there was one. Meanwhile, the two young farmers I sat with to learn about the PLC and ARC programs both grew corn on rented ground but would never qualify for a payment because their farms had no base acres. It was frustrating for them, and by the end of the class they looked defeated. But then again, this wasn't their first rodeo at the government office. Even before the class started, they both knew the likely outcome of our time together.

"He said we are going to understand all the programs clear as day after this hour," one said to the other when we first sat down for the class. The two of them looked at each other, smirks on their faces.

"Yeah, right," the other replied. They both laughed.

Chapter 8

The Cattle Runaround

"THAT'S A GOOD-LOOKING BULL," one of the buyers yelled down from the back row of the sale barn. "Why's he here?"

The local sale barn is housed in a long brick building offset from the road, with large round bales of hay to one side and, on Tuesdays and Saturdays when there is a sale, an array of pickups with livestock trailers in the parking lot. It was a rainy Tuesday in March at the beginning of the COVID-19 crisis when John and I went to see about selling some of our cows and calves. I also wanted to scout the place out, to see for myself how the auction worked and who was there buying cattle. A sign on the door said there would be room for only twenty-five buyers because of the virus, seats assigned according to their bidding numbers—no spectators or sellers allowed. We walked into the main office to talk about selling our cattle, and when we found no one there, John and I made our way up the stairs to the top of the sale ring, to spectate.

The sale area was a mini amphitheater with rows of benches stretching upward in a semicircle. About ten White men sat in the back top rows, heads adorned with baseball caps or straw hats (the kind worn by the local Amish and Mennonite), T-shirts and jeans the uniform of choice. A few of them were on the phone, talking to absentee buyers, we assumed. In the small ring below stood two men behind a hip-high panel holding broomlike sticks, shooing cattle in and out of the small arena, the animals appearing from a door on one side and disappearing out the other. Numbers flashed on a sign above the exit door indicating the size of the animal just sold and the price per pound, and another sign lit up with stats about the cow next up for sale. The auctioneer sat in the middle of it all, in an alcove with a few other men and the only woman in the place, aside from me, recording all the transactions and making sure everyone got paid.

It was a confusing place for an outsider, the frenzied auctioneer-speak hard to follow and bids made with only a slight nod or raise of the finger. Cattle were ushered in and out of the ring in only a minute or two, one arriving immediately after the next, the numbers on the boards meaningless without John's explanation.

There are thirty-four such sale barns and auctions in Iowa,[1] each one replicating a similar process, funneling what amounts to tens of thousands[2] of feeder cattle each week into feedlots, backgrounding facilities (places that feed cattle after they are weaned but before they are big enough for the feedlot), and, occasionally, other farms. Sometimes breeding cows, cow-calf pairs, or bulls also come up for sale. It was the latter that prompted the guy in the back row to yell down, asking why a particular bull was at the auction.

The auctioneer and the others around him looked at one another, startled by the sudden public conversation in the middle of the otherwise speedy and regimented event. "This one's for disposition," replied one of the men. "Don't know about the others." John translated: the bull's semen is good, but he's mean. "He'll make a great bull out in pasture," the auctioneer offered. Translation: don't coop him up in a lot or he could kill you. John continued to explain other elements of the auction obvious to insiders but mysterious to the uninitiated: black cattle go for more money (black = Angus breed = more desirable); single cows usually go straight to the feedlot because something is likely wrong with them (cows are herd animals); and when a buyer calls out "Doc!" the animal is sent to the resident veterinarian, usually to get "preg checked" to see if she is carrying a calf.

Moments after the reply and yet another indecipherable nod, the auctioneer hit his gavel. "Sold."

~

Unlike pigs or chickens, cattle spend their first nine months or so at ranches all over the country in cow-calf operations. John's dad, Leroy, operated that way, keeping a permanent herd of anywhere from twenty-five to fifty cows and two bulls (you need only one bull to impregnate up to twenty-five females), each cow producing a calf in the spring. These cow-calf pairs graze on pasture, the cows eating grass and the calves drinking their milk, until the young ones are separated and weaned at six months of age. At that point, Leroy would bring them into the barn lot and fatten them up for a few months on corn that he had grown. Then, like millions of cattle in the country, these feeder calves were sold at a sale barn, usually to feedlot buyers who "finish" cattle on a fast-fattening diet with nothing green on which to stand or snack. In 2019, this system resulted in more than twenty-seven billion pounds of beef being processed in the United States.[3]

And yet, as I quickly observed on our own farm, cattle do a lot of standing around no matter where they are. "What is the difference if they stand around on a grassy pasture or on a hill of manure in a feedlot?" I asked John one day. It seemed very similar to me, especially for an animal that doesn't seem to know or care where it is standing.

"It's like with your kids," he told me. "They might be perfectly happy lying on the couch all the time eating potato chips, but that is not a diet or lifestyle that is healthy for them or benefits society in any way." I got it immediately.

~

The cattle system is a bit more independent and free than the poultry and hog worlds (which I will explain a bit more in chapter 9): farmers[4] own their own animals, raise them as they like, and don't have to go into massive debt building barns to corporate specs. But because the industry is also highly consolidated, much as it is with pork and chicken, in the end there are very few buyers for cattle, which makes it a buyer's market, often resulting in pitifully low prices.

When Leroy sold cattle back in the 1960s, customers spent close to 20 percent of their income on food, and farmers received almost one-third of that.[5] Today, although Americans' incomes are much higher, they spend only 10 percent of it on food, and farmers receive eight cents for every dollar spent.[6] Even though a cow spends most of its life on the pasture, cared for by the farmer, it is the sale barn, feedlot, processor, and, most of all, the brand under which it is sold that make the other ninety-two cents.

As we drove away from the auction, I was left wondering why anyone would sell cattle at a sale barn. The process leaves no opportunity for sellers to differentiate their product—animals that enter the ring do so with little information attached, no details about how or where they lived or even what their genetics may be. It is an auction in which, during a pandemic, at least, not even sellers witness who buys what and why. The only clues buyers have about the health or vitality of the cows come from the two minutes they are in the ring and perhaps the reputation of the sellers, if they are lucky enough to have their name recognized.

And yet the sale barn is a whole lot better for many producers than the other options available to them, even in a coronavirus-free world. In some states virtually no "fat cattle," those ready for slaughter, are sold at auction. Instead producers and feedlots alike must take what the one and only buyer in town offers, ultimately channeling their cattle to one of the "big four" meat processors—Tyson Foods, JBS, Cargill, and National Beef Packing Company (JBS and National Beef are both Brazilian owned)—which now control an estimated 80 percent of the beef market.[7] There is no competition and therefore no free market to help moderate prices and supply, just ranchers who have to maximize their efficiency, raising as many cattle as possible to the heaviest weights, to eke out enough profit to stay alive. And now American ranchers are in competition not only with their neighbors but also with other cattle raisers around the world: country of origin rules currently allow processors to label meat as *from* the United States even if it was only repackaged here before heading to the grocery store. At the end of the day, consumers worldwide buy from a pipeline in which no one knows where any of the beef was actually raised or by whom, at prices the big four decide they are willing to pay.

The #faircattlemarkets campaign started on Twitter in the fall of 2019, alleging that not only are the prices ranchers receive too low, but also feedlots are losing money—on the order of $200 per cow—while the big four make an estimated $500 per head,[8] totaling a profit of $1 billion a month. "Farmers and feeders come out here and bid aggressively at auction" to own the cows, said sale barn owner Shane Kaczor. "But by the time they take them back to eastern Nebraska or Iowa and [the cows] are fat, they can't hardly get anyone to come look at them. No bidding takes place—they are told what they are going to get for them."[9]

~

A few weeks after our sale barn visit, John and I started wrapping our brains around how and when we might sell some of our grass-finished cattle. They were more than a year old by then, older than John's dad would have sold them in the past, and they, and the land, were looking great. Even though grass-finished cattle usually take eighteen to twenty-four months to fatten, the fact that we were grazing them on the best forage meant that several of them were getting pudgy on the earlier side of that time period, and some would be ready for market.

In terms of environmental impact, our grass-finished beef was as good as agriculture gets. Southern Iowa is naturally green and lush from March to November, and, in contrast with California or Nebraska, no irrigation is used on farms in this area, ever. In our neck of the woods, water is typically abundant, falling from the sky year-round. John and I also never intended to use chemicals or synthetic fertilizers, or administer antibiotics unless an animal was sick, or feed the cattle anything but what grew on the pasture. We didn't clear trees to create rolling pastures and decided not to tile fields (drain them with underground pipes into nearby creeks) in the future—the landscape could remain as nature intended while we continued to produce food, an invasive act no matter what precautions we took. Like the bison, elk, and antelope that once grazed on the plains, eating only a bite or two of grass before moving on, our cows were moved every day, allowing the grass to regenerate and grow deeper roots, some say, sinking carbon into the ground. If there is such a thing, south central Iowa is nirvana for pasturing cows, where woods protect them from heat, rain, and wind and they drink rainwater collected in ponds and streams.

Yet to be clear, beef is still beef, and the rate at which Americans, and an increasing number of people around the world, eat it and other meats is not sustainable. Cattle emit impressive amounts of methane into the atmosphere—they burp and fart a lot—and in order to raise large numbers of animals, ranchers move massive numbers of them through the system quickly by fattening them on grain, which requires valuable water and nutrients, chemical fertilizers and antibiotics.

John and I both feel that all meat—be it beef or chicken, pork or fish—should be consumed sparingly. The simple mathematical fact is that if humans raise livestock only on pasture, without irrigation or grain, there will not be enough for everyone to eat meat every day, let alone at every meal, as many Americans have made a habit of. (So eat your vegetables.)

And although environmentally speaking, grass-finishing cattle is ideal, there are also downsides to it for ranchers. It turns out that moving two herds of cattle every day—we separated yearlings from mother cows before the new calves arrived—along with the necessary fences takes a surprising amount of time. The bulls must also be kept separate from the ladies until we want them to do the deed, and figuring out where all of them can access water and shade in every paddock makes for some fancy footwork. The cost of fence, while not all that expensive per foot, can get out of hand when you have multiple herds or want to keep the cows out of things like ponds, which all adds to the cost of your grass-finished beef.

Back in January 2020, before our trip to the sale barn,

we had met a buyer who was interested in our cattle for his organic grass-fed brand, as long as we had started the organic process. He would be able to take six to eight calves, he told us, and could offer us at least fifty cents a pound more than the market price for commodity beef.

This sounded okay back then. But while the price of beef in the supermarket rose steadily throughout the COVID-19 pandemic as consumers feared a lack of supply and the big four took advantage of the moment, the price of live cattle continued to nose-dive. Some sources reported processors making up to $600 per head in March 2020[10] as ranchers experienced increasing losses.

After meatpacking plants became the center of coronavirus outbreaks, the big four slowed processing (although they made few changes that activists and workers asked for to make the workplace truly safer), sending a ripple effect down the meat supply chain to the growers. Producers and feedlots were stuck with the animals they had planned to sell, the supply far outstripping the big four's ability (or desire?) to process it.

It was a buyer's market, with millions of head of cattle of the right weight ready to be bought. The glut in the market—as we have seen with all commodities—pushed prices down even more than the pathetic pre-COVID 19 rates. So the fifty cents we were offered above market price could in fact be terrible when it came time to sell. Fifty cents more than nothing is still pretty dang close to nothing, we quickly learned, especially in a market where *consumers* are hungry, arguably too hungry, for local beef, buying it like hotcakes.

A light bulb went on in my head. If demand for beef is so high during the pandemic and we have some cows ready—and other cull cows (older cattle ready to "retire") to move out of the herd—why not direct market our beef now? Maybe we could sell five or six cows to the tiny group of friends we'd amassed over the past year in Iowa, I thought, and ship beef to others who live farther away. We should take advantage of the market, it seemed, and make the farm some much-needed income sooner rather than later.

We emailed Story City Locker—a processor about two hours away, known at the time for small-scale processing, with Animal Welfare Certification and one of the only organic facilities in Iowa. Could they refer us to a place nearby where we could have our cattle processed in the next few weeks to sell to people in the area? I envisioned us busting into the grass-finished meat market with hundreds of pounds of beef for sale in just a few weeks, at a price point above what the fifty-cent man had offered but well below the $5 per pound for ground beef I'd heard a woman complain about in the supermarket a few days earlier.

It was the end of March 2020, early in the pandemic, when John opened the email response from the owner of Story City. The last, and most important, line of the email was shocking. "If you go to this link you can find lockers in your area. All are booked up, so plan ahead for 2021."

2021? That was at least ten months away.

~

The rules and regulations for meat processing in the United States are yet another complicated system farmers outside of mainstream agriculture have to learn to navigate. Federally inspected meat processing in Iowa is mostly owned by large multinational companies such as Cargill and Hormel, and these are not places where ranchers like little ol' us can bring our herd. What we needed was a 1ACB processor, one licensed by the state to process meat, which would then allow us to legally resell to customers in the state of Iowa.

We called the only 1ACB facility within an hour-and-a-half drive. "We are totally booked and are not doing anything that needs inspection," the owner explained. "It just takes more time, and I don't need to do it now with so much business coming in. Good luck getting in anywhere," he added.

On to plan B. There are also custom meat lockers where you might take deer or other hunted wild animals (including buffalo, which, unlike cattle, no one wants to have walk into a slaughter facility alive and can therefore be killed on-farm and brought in to be butchered). Custom butchering meant that we would need to line up customers to sell them the *live* cow. Legally speaking, we would then just be dropping off their cow at the facility, where the customers would later pick up the meat, cut as they wanted.

Maybe custom was better anyway, I rationalized, because then we wouldn't have to sell our cattle one rib eye and pound of ground beef at a time. We wouldn't need certified freezers or a place to store massive amounts of meat. We could just hand over the whole animal and have the customer deal directly with the processor about how they wanted it cut.

After two more "No, we are full" responses from custom shops, we finally got someone on the line who told us to leave our name and number for "one of the ladies" to call back. Two days later, after no one called us back, we tried again. "We are not booking any appointments until October 2020. Call back then and we will take appointments for next March."

It was mind-blowing to think about how much meat is currently processed through the system in the United States when we couldn't secure a date for a single animal's ending. Currently, every American consumes almost two pounds of chicken *per week* (ninety-six pounds of chicken per year), one-third more than we did in the 1980s. In order to put chicken into every Caesar salad and noodle soup in the nation, contracted farmers raise more than nine *billion* birds each year in confinement operations, an impossible feat for small-scale farmers to accomplish with slower time lines and feeding regimes. JBS, one of the huge meat processors, which processes ninety-two thousand hogs daily and more than forty-five million chickens per week,[11] produces "6 million 8-oz. servings of pork per day"[12] at its location in Ottumwa, Iowa, near the farm.

It is likely that the people I heard complaining about the $5 per pound ground beef in Ottumwa actually worked for JBS, one of the companies that increased prices during the pandemic while temporarily closing three processing facilities. JBS employs more than 2,200 people,[13] or close to 10 percent of the population of the town, where per capita average income was $22,851 in 2018.[14] Meanwhile, in 2018 William Lovette, the former chief executive officer of JBS, made $7.4 million in compensation, the equivalent of 324 average earners in Ottumwa. The company reported "consolidated revenue" of $629 million in just the second quarter of 2020.[15]

In rural economies gutted by a decreasing number of farms and a loss of manufacturing jobs, towns such as Ottumwa have grown reliant upon food processing as one of the only industries available for residents. And yet paying employees low salaries creates a vicious cycle—national food policy supports growers in overproducing commodities sold for rock-bottom prices, which then are processed by low-paid employees, in order to keep the prices low and the percentage of our income spent on food the lowest in the world. The company—and its shareholders—are the ones who make the profits, while the government shovels out money to keep the farmers on the land. A curious system, to say the least.

~

"How does March 23, 2021, sound?" Finally, we got someone at a meat locker on the phone who would make an appointment with us for a few animals. A year from now we would have our chance to make some money for all our hard work, much more money, it turned out, than if we sold them into the feedlot system. Had we sold our animals at the sale barn (as of January 14, 2021), our top-of-the-line heifers would have gone for about $1,200 per animal. Yet if we sold them directly to consumers, cutting out the intermediaries who pocket the profits in the feedlot system, that number could jump closer to $3,000 gross for each animal. Ten animals would net us about $22,000 after our costs were factored in, for two years' work. But if that amount was added to hay sales, sales of some more cows to wholesale beef companies (in the grass-finished market, not to the feedlots), and custom grazing in which we graze other people's cattle for a fee, we could possibly make enough to keep the farm afloat, and perhaps go on a vacation to somewhere not so exotic once in a while.

Certainly, our profits would not come anywhere close to the millions made by CEOs, compensated stunningly for holding meetings in air-conditioned offices and flying to meet other important people on private planes. But this appointment at a state-inspected meat locker would allow us to keep raising high-quality beef while regenerating the land and improving the soil and forests.

We took it.

Chapter 9

Keeping Up with the Joneses

LESLIE MILLER'S CORNER OFFICE AT Marion County Bank looks out onto a square that is replicated in almost every Iowa town: a grand government building in the middle of the block, with sleepy, often unoccupied shops and nonprofits, restaurants and church offices lining the four sides. I had come to Miller's office because I wanted to understand more about farm financing, how her bank decides who gets loans and for what types of ventures. Having worked in agricultural lending since 1976, Miller has watched trends come and go and technologies evolve, and she has experienced more than one wave of farm foreclosures; I was curious about her take on the current state of agriculture.

We started out discussing the financial needs of farms, bank policies, and the pros and cons of local banks. But almost immediately our conversation turned to an expensive agricultural phenomenon that has become one of Iowa's largest industries: hog confinement.

Miller explained how it works. In this corner of the world, most hog facilities are built to house 2,490 pigs, staying below the 2,500 magic number that kicks in environmental regulations.[1] A facility with 2,500 hogs, for example, must be more than 1,800 feet from a neighbor's house and 2,500 feet away from a park, but with fewer than that number of pigs, growers can put the buildings virtually anywhere.[2]

Farmers first find a company with which they will contract each year to sell the hogs. With the promise of an agreement, farmers then pay for the cost of erecting and maintaining the building, borrowing 80 percent or so of the more than $750,000 price tag to build it to the company's exact specifications. A contract with a hog confinement group such as Iowa Select Farms, Prestage Farms, or Cactus Family Farms is signed, ensuring the producer will be paid for the number of "pig spaces" within the unit whether the unit is filled to capacity or not. The pigs then arrive, owned by the company, not the farmer, to be raised exactly as the company says. When the hogs reach the desired weight, they are sent to processing, usually at one of the huge processors, such as JBS or Tyson Foods, and the process starts over again for the farmer. A typical contract is for ten to twelve years.

"The odd thing in my mind with this whole process," explained Miller, "is that you are basically borrowing $700,000 to buy yourself an income, and you really don't make that much." Miller calculated that for a ten-year loan for $750,000, a farmer would pay almost $8,000 a month[3] while also managing hogs seven days a week, twenty-four hours a day. According to Cactus Family Farms' website, its payout to farmers of $44 per pig space means a "potential additional on-farm income [of] over $20,000 per year."[4]

So why would a farmer erect a hog house? According to Miller and others I talked to, there are essentially three, and only three, reasons to do so.

First and foremost, you would need to already be a corn and soybean farmer, looking to diversify your farm and bring in new income on land you already own.

Second, because you already grow corn and soybeans, your farm can use a lot of the manure generated by the hogs—thereby saving thousands of dollars each year in fertilizer. Cactus Family Farms' website estimates that the manure created from the hog house is equivalent to $18,000 worth of fertilizer.[5] But if the farm does not have land on which to spread it, manure can be a very costly issue. With each hog generating 1.3 gallons of poop each day,[6] keeping 2,400 pigs in a confinement facility will mean finding a place to dump almost 94,000 gallons of pig shit each month, the equivalent of one and a half Olympic-size swimming pools full of pig crap each year.

Third, most families decide to go into hog confinement because more than one generation needs an income from the farm. If John and I had been able to return to Iowa twenty years ago, when we were in our thirties, but Leroy was still going strong and needed his own income from the land, there would have been very few ways to involve us in the operation *and* make additional income. It doesn't take a lot of land to put up a hog house (or a poultry facility, which is similar), so confinement offers an attractive option for families to make more money on their family's land.

"But a lot of times, that is really for the benefits," Miller added, also citing the $20,000 per year a family might need to pay out of pocket for health insurance,[7] which is, oddly, about the same amount a hog facility would net. And, Miller added, if a farmer has to buy water for the facility, or dispose of the manure elsewhere, or if someone in the family gets sick, those costs can make that slim amount of additional revenue quickly shrink even smaller.

Miller told me that in the thirty or so years hog confinement has been in Iowa, she has seen only a handful of cases in which the deal did not work out the way the farmer was told it would. The family usually makes what the contract says it will, and if the building is kept up, it can be used long after the initial ten-year contract is completed. That's when there is some real money to be made, she reported, when the facility is owned free and clear, if all goes according to plan and the farmer has invested in upkeep of the building over time.

Even though there is no part of me that wants to run a hog confinement facility, I can see the allure. For all the smell, the sleepless nights full of worry about debt, and the questionable humaneness of confining animals as smart as pigs,[8] at least these units offer a way for a family to make *some* guaranteed income. Of course, that financial security comes with a trade-off, most importantly a total loss of independence—that highly coveted but elusive holy grail for farmers.

~

Borrowing money to farm in the United States, of course, did not start with the advent of hog confinement. Even British colonists (at least the White men) looking to make a buck growing cash crops in the early days of our nation took out loans to buy land and slaves.

"The US has always been the breadbasket of the world, and there has always been a huge export market," explained Professor Sharon Murphy of Providence College in Rhode Island. Murphy, a business historian and an expert on early American banking,[9] explained to me that unofficial small, local banks popped up all over the young states in the 1810s to lend money, mostly for agriculture.

But in 1815, a chain of events began that led to the nation's first financial crash, a scenario that, as a nation, we have repeated again and again in our history. It started when the British purchased a huge amount of grain and cotton from the nascent United States after the largest volcanic explosion in recorded history changed the climate in Europe for more than a year, devastating crop production.[10] It was a sale the United States desperately needed to pay off debts incurred during the Revolutionary War, and the boon soon drove up commodity prices. "As demand for these goods starts spiking," explained Murphy, "people start saying, 'Oh, look at that! If I expand my farm and plant more, I am going to make a killing here.'" So, Murphy continued, farmers took out more loans and went further into debt, while the banks grew less scrupulous about who they lent to and how much they loaned (a scenario, Murphy emphasized, repeated most recently in the housing crisis of 2008).

Then, as world politics shifted, exports came to a screeching halt and farmers were left sitting with the goods, now far in excess of what they could sell. The price of cotton plummeted. But instead of cutting back on planting in order to create *less* supply, farmers again doubled down, borrowing more money to expand and increase production in hope of wringing out any tiny profit they could. As should have been expected, instead the prices fell further and farmers were unable to pay the money due.

"You have all these people who are overdrawn [at the bank] . . . and they are left with debts they can't repay," Murphy told me. "People start reneging on their loans, and all of a sudden, the banks are underwater. You have the cascading effect of banks folding and people panicking." The price of land crashed as families were unable to pay back their loans, sending the entire market into a tailspin. Prices in Pennsylvania, for example, went from $150 per acre in 1815 to $35 in 1819.[11]

Sadly, the lessons of 1819 were not learned, and the scenario has repeated itself again and again throughout history. National panics due to unscrupulous bank lending, overborrowing, land speculation, and overproduction (often with the help of railroad companies) brought on recessions in 1837, 1873, and 1893. And by the start of the

Great Depression in 1929, farmers overborrowed not only for land but also for bigger plows and newly invented tractors. The Model A by John Deere, for example, was one of the first mass-produced tractors, put out in 1934 during the height of the Depression.[12] Better machinery meant increasingly efficient production, and while the abundance of crops grew, so too did debt. Almost two million farms—coincidently the same number of farms we have today—were foreclosed on in the period from 1921 to 1940.[13]

"Often what is best for the individual farmer is counter to what is best for the industry as a whole," Murphy said, concluding our phone call. "The individual farmer, in order to survive in the system, needs to buy these tools, which are also expensive and cause them to go into more debt. So individually, buying machines is the best decision for them. But as an industry, everyone is contributing more supply when the problem is already oversupply."

Because many farmers fear that their neighbors will grow higher yields and flood the market—thereby making the price of each bushel lower—they too must produce as much as possible in order to cover their own costs. When there is a good year, it seems like a great time to get ahead and buy that bigger tractor, increasing your potential for even higher yields in the future. When it is a bad year, it's suddenly imperative to have that bigger tractor just to stay in the game. And so the cycle continues.

~

As Leroy tells it, when hog confinement began in Iowa, many farmers knew full well the irreparable damage the practice would have on the industry and were pissed off about it.

"Before, bankers would call hogs 'mortgage lifters,'" Leroy explained as we sat at his kitchen table. "When a farmer raised a big batch of hogs and sold them, he had a big wad of money and could pay off his mortgage." Raising hogs—slowly, on pasture—was reliable and consistent income.

But then, in the late 1970s, someone in town had the idea of going in together to buy bigger, more intensive facilities. Leroy and about eight other farmers and non-farmer investors decided to go into business, creating MoCo 10, short for "the Monroe County ten," although, Leroy added, there never were ten people involved. The group borrowed $750,000, the equivalent of $3.3 million today, hired a manager and an assistant, bought breeding stock from Wisconsin, and started raising hogs.

"But as we started producing hogs, people could see the handwriting on the wall, and they were really angry with us. There was going to be *a lot* of hogs on the market, and people could see that was the end of the lucrative hog market," Leroy explained. "There would be so many pigs out there, there wouldn't be demand for them. Which is exactly what happened."

For Leroy, an early adopter, the situation worked out okay. The money did roll in for quite a while, but by the late 1980s—during and after the farm crisis—more confinement units went up across the landscape as farmers looked for ways to improve their income.[14] An article on the early days of confinement contracts in Iowa in *Success-*

ful Farming magazine included a quote from Al Bormann, a farmer whose family weathered the 1980s crisis. "For a lot of people, it was their only way out," said Bormann, referring to the money guaranteed to farmers when they signed contracts with companies to raise hogs. "We realized . . . people would not be here if it wasn't for contract feeding, and the community would not be here if all those people left. We saw the incredible number of people who left Kossuth County in the 1980s. It would have been worse. [Contracting] offered people a way to stay."[15]

The confinement units MoCo 10 put up were tiny in comparison with these newer, bigger facilities, and the group realized they too would have a hard time competing. MoCo 10 sold the venture and disbanded in 1993, just before the market crashed in 1995. That year hogs went for $8 per head; they cost $75 to raise.

~

There have been attempts throughout the country's history to limit the supply of commodities and to guarantee farmers the cost of production, a concept also known as parity. Franklin Delano Roosevelt stabilized prices during the Great Depression by stockpiling commodities and paying farmers to grow less through the Agricultural Adjustment Act in 1933. The idea was to offer financial incentives for farmers to limit their production and for the nation to store the excess grain until such time as the price rose again. The idea worked, and many farmers, including Leroy, stored grain on their farms as part of the deal.

But by the 1960s storage facilities full of grain were

costing taxpayers millions of dollars, and the politics of the day demanded change. Companies including Cargill, the largest privately held company in the United States, lobbied for the end of supply controls and the higher prices that came with them. "For Cargill and the grain trading companies, this was anathema," wrote Wayne G. Broehl Jr., referring to the company's stance against mandatory limits proposed in the 1961 Farm Bill. "Their business was built on trading, and it would not be too much of a simplification to say 'the more [grain produced] the better.'"[16]

Unsure of what to do to keep prices stable and farmers on the land, the John F. Kennedy administration put the question to the farmers, calling for a "wheat referendum." If two-thirds of wheat producers voted for it, the amount of wheat grown would be limited, thereby restricting the supply and driving up prices. But Cargill and other corporate interests, with the help of the American Farm Bureau Federation, a "farmer-member" lobbying group, set out to convince farmers that limiting supply would be bad. "The Farm Bureau charged that the farmers were being forced to surrender their very freedom," one of the bureau's leaflets shouted. "The real issue in the Wheat Referendum is . . . [the] freedom to farm."[17]

The plan worked. A huge majority of wheat farmers voted against limiting supply, deciding that their freedom to choose what to plant—and how much of it—was more important than the price they received in the marketplace. The myth of independence and freedom held more allure for farmers than even the prospect of making more money.

~

The legacy of that decision has not only meant that farmers no longer receive parity for their goods; it has also indebted farmers in ways that are often overlooked. The "freedom" to buy expensive machines to overproduce commodities put farmers on an agricultural treadmill permanently. But during the Kennedy administration, agricultural economist Willard Cochrane also foresaw that efficiencies such as machinery, seeds, and hog houses would mean that the power to make decisions would slowly pass from the farmers into the hands of the nonfarm entities that supplied the money. "For Cochrane," wrote economist Richard A. Levins, "the Trojan horse of better farming through technology was filled . . . with nonfarm investors who would dictate practices on operations that for all appearances looked like family farms. . . ."

Instead, "commercial nonfarm sources: insurance companies, feed companies, processors, and retail organizations" would step in, "and operational control invariably follows financial control. We could thus in 2000 have family farms in form but not in spirit . . . managed by and with their financial risks assumed by a nonfarm organization."[18]

This form of control is everywhere in agriculture today. John Deere now not only sells machines but also lends farmers the money to purchase them; the company more than doubled its net income from financial services in 2018 to $942 million.[19] And while farmers can pay upward of $700,000 for a combine, they actually have only an "implied license" to *use* the machine, says the company: they don't really *own* it and are forbidden to alter or fix the software—which, as in most cars today, runs the entire vehicle.[20] John Deere also finances not only its own products but also products from brands including Bayer Crop Science and Syngenta, offering low- or no-interest loans to buy patented seeds, which farmers would need a combine to harvest.

Today, the hog facility phenomena Leslie Miller described to me also clearly illustrates how Cochrane's biggest fears have been realized. "Family-owned" Prestage Farms and Cactus Family Farms[21] (Cactus sells the contracted 750,000 hogs it oversees to Tyson), for example, may help stabilize income for farmers by guaranteeing contracts. But farmers have given up *all* freedom to decide how and where to raise the animals, making farmers landlords or, worse, hired hands on their own farms.

Rolling Stone magazine in 2018 published an article about the purchase of Smithfield Foods—one of the largest pork producers in the world—by WH Group, a Chinese company: "The company now owns the hogs, the most lucrative part of the business, while the North Carolina farmers own the shit—and all the environmental and human liabilities from it."[22] In 2020, Smithfield was by far the largest "pork powerhouse," according to *Successful Farming* magazine, owning 915,000 sows in the United States (which then produce oodles of piglets for market).[23]

But that's not the whole story. While Smithfield has embraced a vertically integrated business model, owning the hogs as they age through the system, the other huge pork processors have largely shielded themselves from the risk of owning the animals. Iowa Select Farms, for example, raises three times as many sows as Tyson does;[24] Tyson purchases the bulk of its hogs from confinement operators such as Iowa Select. This means that during the coronavirus pandemic, many big meat processors could slow down or stop processing and not lose a dime. They just shut down the pipeline and let everyone else take the hit.

~

But things are better today than they were in the 1980s (or in the 1920s, or in 1819). Right?

Economists often like to talk about how the amount of debt farmers carry today is not like it was during the 1980s farm crisis. Back then interest rates went sky-high, economist John Newton of the American Farm Bureau Federation told me, whereas today's interest rates have remained low. In May 2021 you could get a loan for a little over 4 percent interest, far less than the 21 percent[25] it went up to in the 1980s. In the 1980s, those high interest rates made the land market suddenly stall out and land values crashed, skewing a farm's debt-to-asset ratio: farms owed far more than their assets were actually worth. By comparison, today's land values seem to be more stable and a family's assets are protected—the debt-to-asset ratio is low—making today's economic situation dramatically different.

"But to me, that [metric] is not as meaningful as when you compare the debt to a farm's *income*," Leslie Miller, the banker, explained to me. "Farms just don't make that much money to pay off the loans." Bank loans are based on two main things: a family's entire income, not just the income from the farm (more on that in the next chapter), and the value of the farm's assets—the land and machinery. But if the income of the farm business doesn't cover the cost of the loan, what good is it to buy a combine if the only way you can afford to pay it off is to sell the land you use it on? In 2019 total farm debt hit a record high of $416 billion, and yet total net sales were only $88 billion, 40 percent of which came from government payments.[26]

~

Interestingly, many farmers are willing to accept these concessions—not owning the animals, seeds, or machines they care for; erecting expensive facilities to house operations they do not have the right to make decisions about; and borrowing excessive amounts of money from nonfarm entities to pay for these endeavors using their own land as collateral—in order to continue to work on the land. To me the irony is that although Leroy and millions of other farmers want the freedom *not* to be told what to grow or how to plant, the capital needed to stay in the farming game has left many signing away those very same freedoms simply to keep the farm.

At the end of one of my discussions with Leroy I asked him about this paradox. If farmers are such an independent lot, if freedom of choice means so much to them, how have they accepted the terms dictated by the John Deeres and JBSs of the world?

"Things are always evolving," he reminded me. "You

don't have the money to put up the facility and so you basically agree to become a hired hand on your own farm." With each generation the dream of freedom for farmers has grown more faint.

"It creeps in," Leroy added. "It doesn't happen overnight."

Chapter 10

Everybody Does It—off the Farm

"I WAS SHOPPING FOR A GIFT FOR A FRIEND on Etsy," Margaret Jodlowski told me over the phone. "And I saw a sign that said, 'Behind every successful farmer is a woman that works in town,'" she said with a chuckle. "I thought—is that true? That would be a great research question." Jodlowski was looking for a topic to study for her PhD work (she is now a professor in the Department of Agricultural, Environmental, and Development Economics at Ohio State University), and she decided to investigate it.

For John and I, it wasn't my professor gig that was paying the farm bills. It had been over a year since I left my job in San Francisco and we sold our house and jumped into farming with both feet. By the time I talked with Jodlowski, we had already paid our second round of rent ($13,000), made a second cattle payment of $10,000, and fixed the tractors several times. With a loan from the bank we bought a $10,000 four-wheeler to get around the farm—which saved hours in walking everywhere—and we were also sinking a lot of money into remodeling the hundred-year-old farmhouse.

But I just couldn't wrap my brain around how, in fact, farms stayed in business as long as they did. If we had all these costs while attempting to keep our business really lean, not spending much on machinery or improvements, how did farms that grew corn and borrowed lots of money every year survive?

The answer: off-farm jobs.

~

In 1930 one-third of farm operators worked off the farm part-time; by 2000, 93 percent of farms relied on off-farm jobs for at least part of their income.[1] Today off-farm jobs account for almost *all* of a farm household's income, as seen in figure 10.1.[2]

To clarify the graphic for those of us who have not taken a math class in eons, "median" means the middle, not the average. Half of all farms, then—one million of the total two million farms—will make less than $495 in 2021 (in 2019 the median farm income was even less, $296, as I mentioned in chapter 6).[3] The reason why the chart shows any positive farm income at all in 2019 and 2020 (although the number is so small it is almost impossible to see on the graphic) was not that farmers were more competitive, efficient, or profitable than in previous years; it was that farms received larger government payments than ever before, as shown in figure 10.2.

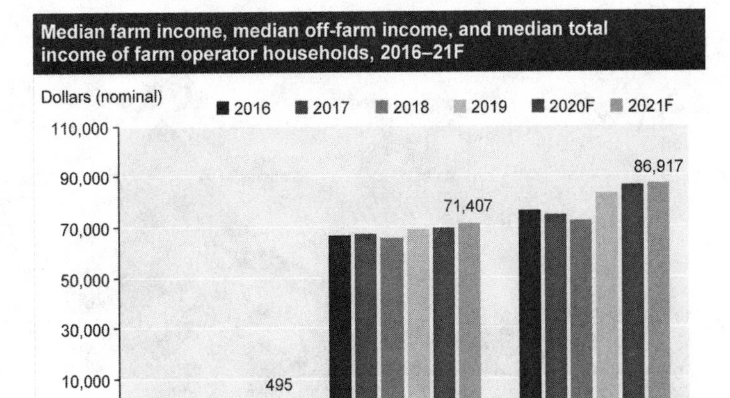

Note: F = forecast. The median is the income level at which half of all households have lower incomes and half have higher incomes. Because farm and off-farm income are not distributed identically for every farm, median total income will generally not equal the sum of median off-farm and median farm income.
Source: USDA, Economic Research Service and National Agricultural Statistics Service, Agricultural Resource Management Survey. Data as of February 5, 2021.

Figure 10.1

"It is interesting that a majority of farm household income comes from off-farm, but that it is not publicized," Professor Jodlowski explained during our call. Her research since the Etsy epiphany has focused on the family dynamics involved in decisions about who works on and off the farm. "There's a social pressure around farming being profitable, a sense that people want to be successful enough not to work off-farm, and there is a kind of stigma around it. It plays into gender norms."

As many of us learned in school, women and work became a thing around World War II, when electric stoves, dishwashers, and washing machines made the drudgery of housekeeping less time-consuming, enabling women to

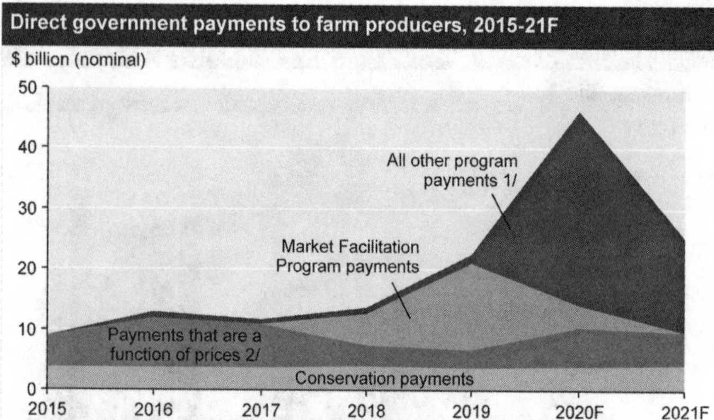

Note: F = forecast. 1/ "All other program payments" includes supplemental and ad hoc disaster assistance, which in 2020 and 2021 includes payments from the Coronavirus Food Assistance Program and the Paycheck Protection Program, and in 2021 also includes payments under the Consolidated Appropriations Act, 2021. 2/ Includes Price Loss Coverage, Agriculture Risk Coverage, loan deficiency payments (excluding grazeout payments), marketing loan gains, certificate exchange gains, and dairy payments.
Source: USDA, Economic Research Service, Farm Income and Wealth Statistics.
Data as of February 5, 2021.

Figure 10.2

finally leave the kitchen and go get a job. Women did so in droves; in 1950, 18 million women worked; fifty years later, 66 million did.[4]

American farms too saw a fundamental change when new and improved machinery came on the scene. According to the 1954 Census of Agriculture, there were a million more tractors in that year than there were in 1950,[5] and farms in the 1950s continued to grow larger as a result of "farm consolidation arising because of the improvements in farm technology and efficiency [and] the rapid mechanization of agriculture."[6]

This shift to mechanized labor on the farm is usually discussed in terms of higher efficiency and productivity—

how many more seeds could be planted or rows harvested each hour, leading to higher yields. Yet an important outcome was also that the new genetics, chemical fertilizers, and tractors allowed individuals to farm much more land by themselves. It took twenty-eight hours of labor to work an acre of corn in 1930; by 1960 it took only seven[7] (today it takes a little more than two[8]). Farm labor dropped by almost 80 percent from 1948 to 2011;[9] today a John Deere combine can harvest two bushels of corn in a single second.[10]

But interestingly, growing more with less labor and less land has not meant that farmers make more profit. It has just meant that instead of only struggling to make a living on the farm, farmers and their families—often the women in the family, as Professor Jodlowski pointed out—have more time to work in town to supplement the family's low income.

You can see the gender norms Jodlowski described at play when you walk into almost any bank or business office in our neck of southern Iowa: women sit at all the desks visible from the front door—working as the receptionists, bank tellers, and saleswomen—while the few men at these businesses occupy the back offices reserved for managers and owners. Yet with women making eighty cents on the dollar[11] for the same jobs men do, Jodlowski added, the norm of women leaving the farm to work is actually not ideal for families interested in maximizing their income. If economics were the only factor in deciding who works off-farm, then men—who make more per hour—would be at the desks rather than in the fields.

But, of course, there are many other reasons why women (and men) work in town.

"I enjoy my work; I am happy to work," my friend and fellow farmer Casey Havre told me about her job certifying organic growers in California. Casey also runs the business side of the farm she and her husband, a fourth-generation Central Valley farmer, own, growing eighty acres of almonds, cherries, citrus, and table grapes. Like many women, Casey feels her off-farm job is part of her identity, and she loves meeting other organic farmers in the process. But, she admits, the stress of bringing in all the income for the household can also be overwhelming, particularly in hard times.

"I have independence in my work. I see different people all the time, and it gets me out of the house and out of the office," Casey told me. "But that's not to say I would not love to just stop. But I can't. I have always had an off-farm income because the farm doesn't support two people and the workers we need. Especially in the last six months or so, I have been paying for the farm with my income since COVID-19 hit."

Casey's experience is a common one, especially in hard times. Jodlowski's research found that money made at off-farm jobs is not only used to fund the day-to-day needs of the home, allowing the family to buy such things as food and clothing. The off-farm income is also used to pay for the farm's expenses. As Leslie Miller, the banker, mentioned in the previous chapter, the money lent to buy expensive machinery and high-yielding seeds, for example, is calculated on the basis of the farm's assets and overall *family* income, not on the *farm*'s ability to generate the needed cash flow. The debt owed by a farm is often entirely dependent upon the income the family makes in town; most farms can't function without it. If you think about it, this may be the real reason family farms continue to exist—it takes a family of income earners to help pay for the activities of a single farm.

This is again Willard Cochrane's agricultural treadmill in action, the self-perpetuating cycle that keeps farmers borrowing from banks and seed companies to buy the latest, greatest thing in order to increase their efficiency and produce more. But off-farm work adds yet another interesting component to that cycle. Often it is farmers and their families working at the bank, loaning money for this year's seed, or at the John Deere dealership selling that expensive combine or servicing it when it breaks. Farm family members *are* the managers at the meatpacking plant and the fertilizer spreaders, the government Farm Service agents, and, in Casey's case, the organic certifiers who work in the industry. In a bid to sustain our own farms, we farmers inadvertently keep one another in debt. And, to add insult to injury, businesses such as Chinese-owned Smithfield Foods or even out-of-state certifying organizations end up further draining profits out of rural communities and into central business offices far, far away.

~

Yet the need for cash isn't the only reason in-town jobs are essential; as in most American households, health insurance also plays a key role. As is the case for any small business or independent contractor, if a farm operator gets sick or has an accident, there may be no one else to run the business; without insurance, a fall or a serious diagnosis can bankrupt the enterprise. But farmers are also dependent upon the health of their bodies to do their job. And getting hurt on the farm is unfortunately a real concern, just as it is for those working in other dangerous jobs, such as roofers and truck drivers.[12]

The all-encompassing nature of farming adds a further wrinkle. While many professionals see their jobs as central to their identities, a farm is often the family's home and holds their collective history. In other words, farmers without insurance risk losing *everything* if an accident happens; obtaining health insurance at an off-farm job is in effect a risk management strategy, not only for individual farms but also for our national security.

"It is important for everyone to have health insurance," Alana Knudson, a specialist on rural health at the University of Chicago, told me. "But it is uniquely important for farm and ranch families because of the connection to their livelihoods, to their way of life. And there are also long-term implications. It's in all of our best interests that we have an adequate food supply. It is a national security issue."

~

Yet the time spent at work off the farm—and driving to and from jobs far away—can add a lot more stress to the already intense cycle of farming. Starting a business of any kind takes focused, committed time; if you have to work at another full-time job, you are left with little brainpower to fully engage in your new business. Farming is also a profession that happens mainly in the daylight, limiting the possible hours outside of a nine-to-five job one can spend weeding, tending animals, or trimming trees. And for dairy farmers, doing anything else aside from milking cows or goats two times a day, every day, is not an option.

Many new farmers report feeling that, because they work at another job, they never really get enough time to establish themselves or to take advantage of available resources. Even more experienced farmers such as Casey—who has lived and worked on the farm for more than fifteen years—find that time and financial deficits go hand in hand. Less money means she must spend more time at her off-farm work and less on the farm. "When things are good—then your whole world opens up to the possibilities of what you can do on this beautiful land," she told me, explaining how in better times a few years ago she put in an olive orchard and spent time making olive oil. "But the thing is, when it gets tough, then you don't have time to go out and take care of it. I wish I had more time and more freedom to do those kinds of experimental things. That is the most difficult part for me."

On the one hand, Casey's desire for more time sounds like a common American complaint—who wouldn't like to have more time to do the things they enjoy? But fruit and vegetable farmers such as Casey are not growing food as a recreational pastime; the fruits of her labor are not the same as for someone who desires more time to ski or scrapbook.

Despite the disparagement of small-scale agriculture as irrelevant hobby farms that grow only a fraction of the country's food, the fact is that small and midsize farms make up the vast majority—almost 95 percent—of the nation's agricultural enterprises and account for more than 40 percent of production (see figure 10.3).

This kind of chart is often used by agricultural extension agents, seed companies, and academics to say that bigger farms are better simply because they produce more. If less than 3 percent of all farms produce more than 40 percent of all agricultural output, why not just have fewer, larger farms?

Yet it is important to remember that much of the agricultural output of this nation (depicted in the figure) does not feed humans and is responsible for the overproduction and low prices farmers battle. Many larger farms also receive gargantuan government supports, putting their self-sustaining profitability in question. Plus, gross sales are not an indicator of profit, nor are they an indicator of farm size; smaller farms might make more profit in the end than huge debt-ridden ones. The argument could also be made that, if given the ability to experiment and spend more time on the farm, Casey and so many others like her might figure out ways to become more competitive and profitable in the marketplace. They would likely find time to collaborate, to create desperately needed infrastructure, and to make products more accessible to a wider range of buyers.

As the average age of farmers keeps creeping up and new people are needed to tend the land, the lack of time

Farms and their value of production by farm type, 2019

Percent of U.S. farms or production

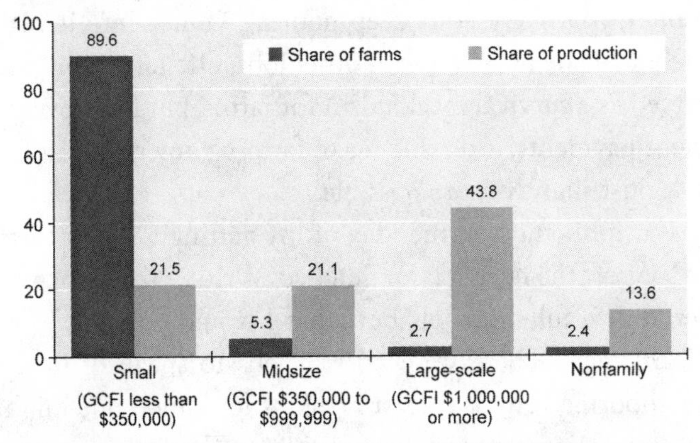

Note: GCFI = annual gross cash farm income before expenses. Nonfamily farms are those where the principal operator and their relatives do not own a majority of the business.
Source: USDA, Economic Research Service and National Agricultural Statistics Service, Agricultural Resource Management Survey. Data as of December 2, 2020.

Figure 10.3

farmers have to hone their craft creates a difficult conundrum. The American public has invested over $175 million in training new farmers over the past decade,[13] teaching those passionate about growing food everything from cropping techniques to best business practices. But all the training in the world won't help if new farmers never have time to dedicate themselves to their farms. Their operations will never develop the resiliency necessary to make it through hard times.

~

I can't think of another industry in the United States that is almost entirely dependent upon other, often unrelated, industries for support in the way agriculture is. Many restaurants are unprofitable but go out of business in a few years; if your family owns a shop that no longer makes money, you likely don't keep pouring money into it with income made by working at other jobs. Off-farm work has served as a survival mechanism for farms, but like government payments, it doesn't make farming any more profitable, just slightly more possible.

For John and me, the idea of my getting a job in town to support the farm felt far-fetched. Writing this book was certainly a full-time job, but what I would do after, I just didn't know. There were no real jobs to speak of in the neighboring towns, at least not jobs for which one might need a master's degree in journalism. Freelance journalism wasn't a job that paid well or had benefits either, and although Central College in Pella, Iowa, is about a thirty-minute drive away and I had taught for years at the University of San Francisco, it wasn't statistically likely that I would ever score a position as a professor again.

The truth is that we could afford to live off the profit from selling the house in San Francisco, at least for the first few years. We had invested a chunk of the money in stocks that, although they are considered socially responsible funds, likely support companies like John Deere or Frito-Lay—the same companies making a lot of money off farmers for their own chief executives and for shareholders like me. While infuriating, this had its own perverse twist—it was another example of the treadmill at work. In this small way we too profited from the system, but at such a cost to so many people.

Chapter 11

A New Narrative

BACK IN THE FALL OF 2019, when John and I first moved to Iowa, we attended a weekend retreat for beginning farmers, mostly of the sustainable agriculture sort. For two days farmers, ranging from twentysomethings with vibrant community-supported agriculture (CSA) memberships to those of us in our fifties moving back to the family ranch, discussed our agricultural hopes and dreams. A workbook started us out talking about our mission on the farm. We took a quality of life assessment, listing our skills and how much time we were willing to spend marketing our products. We talked about the things we most valued in life and how farming would support them.

Late on the second and final day of the retreat we turned to finances. "Does this enterprise have promise to deliver large enough revenue relative to the investment required?" a worksheet asked us on page twenty-one of the twenty-nine-page booklet. Only then were profit and loss statements, cash flow, and balance spreadsheets introduced. After a quick explanation of how to fill out the sheets, we promptly ran out of time, the financial tasks of our farm businesses left yet again to another day.

But the mention of money spurred a conversation the end of the session could not stop. A few working farmers in the crowd brought up how picking veggies early in the morning and heading to the farmers market for ten- to twelve-hour days was not working out financially as they had hoped. A thirty-year old participant who kept meticulous records of the products sold in his local community wondered aloud how many years it would take for him to make a profit. Another brought up how she and her farming friends talk about the fact that they can't afford health insurance and worry they will never be able to put away anything for retirement. "What happens if I want to start a family someday?" she asked rhetorically; the room was silent.

The weekend was a great opportunity to step away from the daily tasks of the farm and think about what we were doing. John and I came away with a sense of how our values impact our decisions and a rough idea about which enterprises might work best for us. Yet even if we had spent more time on learning the financial aspects of farm businesses—skills that are the focus of many other beginning farmer courses—rarely do these classes make mention of how little money most farmers make or of the huge structural challenges that exist in agriculture. As a result, when dealing with all I have outlined in this book, many new farmers come to feel like failures.

"We don't want to scare people out of farming," the workshop leader told me when I asked why we hadn't talked about any of the real challenges people encounter when starting a farm: lack of credit, little access to land, low margins. Maybe he should instead start the workshop by explaining that most farms don't make a profit, I suggested, and then spend the weekend equipping beginning farmers with tools they'd need to beat the odds. He looked at me as if I had asked him why he wasn't teaching about farming on Mars.

"If people know how hard farming is," he told me, "they won't do it."

~

Maybe he was right. Maybe the necessity of attracting new farmers means that we need to excite them about agriculture and cross our fingers that they can overcome the challenges when the hard times inevitably arrive. Less than 10 percent of farmers are under thirty-five years of age, while more than one-third are over sixty-five;[1] compare that with the general workforce in the United States, in which more than thirty percent of workers are in their twenties and baby boomers make up only one-quarter.[2] Sixty percent of land in Iowa is also owned by people over the age of sixty-five.[3] Agriculture needs to attract new blood, and telling people how dire farming is economically may certainly dampen their interest in the profession.

But there is another reason we didn't talk in depth about the problems farms face: they don't fit into our national mythology about farming.

~

Two stories dominate our collective narrative about farming in America, showing up everywhere from advertisements to political speeches, documentary movies to farm journals. These two pervasive myths dictate how we as a nation think about farmers and how farmers themselves perceive their own profession, and are narratives I have alluded to throughout this book.

The first is one I call the "bigger is better" myth, the belief that if a farm is to be successful, it needs to grow bigger. It is a tale that extols the success of American farmers due to increased yields, better machines, and more efficient seeds, techniques valiantly adopted by farmers in order to feed the world or, more likely today, to meet the demands of a growing global, meat-eating, middle class. This myth considers operations like ours hobby farms, recreational pastimes of the wealthy who don't need to make an income. Smaller farms, the narrative tells us, aren't real businesses.

This mindset so dominates commodity agriculture—reinforced repeatedly by reputable sources such as university extensions, the American Farm Bureau Federation, and the US Department of Agriculture—that most farmers never question its validity or wonder about its role in their own farm foreclosures. And, sadly, it is the American public, fed more cheaply than any other population on Earth, that unknowingly reaps the bounty.

A chat I had recently on Twitter illustrates the bigger is better mindset. The initial tweet was an advertisement for a chemical that claimed it would "help you achieve peak yields and beat your competition," a phrase that struck me as absurd. Who is the "competition" farmers need to "beat"? I wondered. And how would beating other farmers make a farmer more successful? I typed my reaction out into the social media abyss and pressed Send. A self-proclaimed ag expert (not the company that ran the initial ad) responded (see figure 11.1).

I will give @rayfashworthjr credit: it is true that I am not an agricultural economist. And the bigger is better story—like every myth—has some element of truth to it. Once per acre costs are spread out thinly enough, sheer volume can help make up for the cost of doing business. Large farms (which are mostly still considered family farms) can also weather bad years more easily than smaller ones because of their larger reserves. Which makes some, though not all, of the country's largest farms profitable (although, since they also receive the lion's share of government support, how profitable are they, really?).

But the belief that farmers must compete with one another and that oversupply is a short-term problem is—as I have illustrated previously—part of the reason farmers aren't financially stable today and have rarely been throughout US history. Overproduction is cyclical and, today, inevitable.

The bigger is better narrative also ignores other key realities facing farmers. First and foremost is the fact that most farms are nowhere near the size necessary to profit under the bigger is better model; almost 90 percent of farms gross under $350,000 per year, including most commodity farms. Growing your farm steadily over decades doesn't give you the benefits of a massive corporate

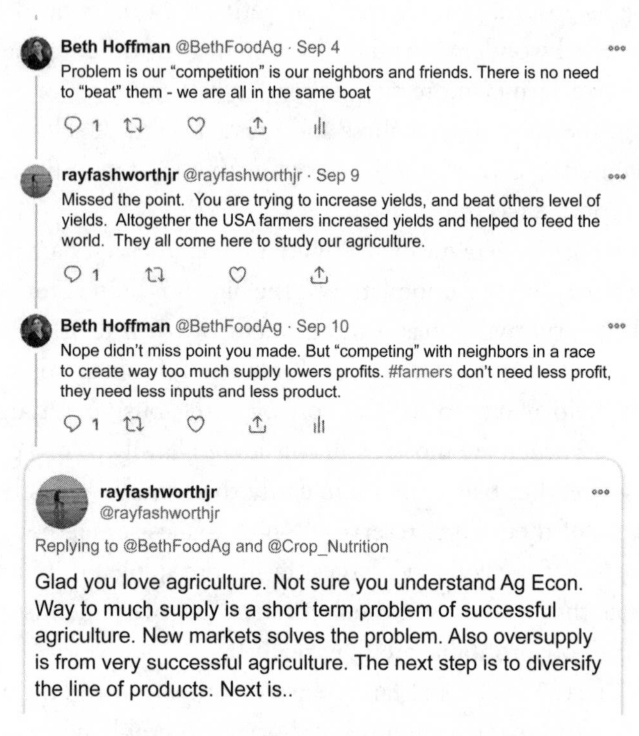

Figure 11.1

operation; getting big is not an incremental game but an exponential one. One more machine and a little more debt are not enough to improve a farm's financial prospects. Instead, to get big you have to go really big: risking it all, working with investors (be they banks or agribusinesses), and buying the latest and greatest equipment and seed year after year to keep yourself relevant.

And even when farmers spend the money to grow large enough that an economy of scale seems in sight, it turns out getting bigger is not actually more profitable. Even in 2010, when corn prices rose, and then stayed high for two and a half years, a study by the US Department of Agriculture found that "producers with the smallest corn enterprises (fewest acres) had the lowest average operating and ownership costs *per acre*, while producers with the largest corn acreage had higher costs per acre."[4] The record-breaking corn yield for 2019, at 616 bushels per acre (the average that year was 168), tended with the very best seeds, sprays, and machines technology could offer, cost $1,400 to grow and netted a profit of less than $1,000 an acre.[5] For a very large 50,000-acre farm—the only type of farm that could afford such inputs—that means a profit of only $50,000 for what would amount to more than 100,000 hours of labor.[6]

In other words, bigger is better takes the kind of commitment, labor, and debt most farmers don't have or want, and even then, most farms don't make very much money, if they make any money at all. It also means that, by definition, many of the farms attempting to get bigger will instead go out of business. The biggest farms need to suck up more of the land, consolidating our nation's greatest resource in the hands of ever fewer people.

A bigger is better mindset also works only if the oversupply indeed stays short-term, as my Twitter friend indicated. Earl Butz, secretary of agriculture under Gerald Ford and one of the biggest cheerleaders in American history of the bigger is better model, famously encouraged farmers to "plant fence row to fence row" in response to the plethora of new markets he saw available to farmers around the globe. And during his years in office, farm income did in fact *increase* by 20 percent in six years, while government subsidies *dropped* from $3.7 billion in 1970 to $500 million in 1976, 13 percent of what it was. Sixty-one million new acres were planted,[7] each acre producing more than it had previously.

But during that same period exports almost doubled, an impossible feat in today's world, in which competition for wheat is extreme from Russia, Ukraine, and Argentina and for corn from Brazil, Ukraine, and Argentina.[8] Corn-based ethanol plants, the "new market" for growers in the 2000s that led to the last big boom in prices, experienced the repercussions of overproduction even before COVID-19,[9] when the industry came screeching to a halt, according to the American Farm Bureau Federation.[10] And in the past twenty years, there have been really only three and a half outstanding years when prices of corn were high;[11] the average price now will not cover the cost of a corn grower's production. Today there are few new markets out there for farmers to tap into, even with products such as corn-based soda bottles and polyester fabric on the horizon.[12] It also goes without saying that when farmers start growing corn to make plastic, the rationale for using some of the best soil on the planet to feed the world vanishes into thin air.

And today those who subscribe to this narrative have, as I've demonstrated in this book, traded away the one thing they valued most: freedom and independence. In a bet to make a profit and stay in business, they gave away their ability to make their own decisions. How they will raise animals, which crops they will plant, how much they will sell their animals for, to whom they will sell livestock: none of it is within their control any longer. Even farmers' right to fix their own tractors is now in question.

But perhaps the largest issue for the bigger is better model is the inevitable disappearance of the rural towns and smaller farms dotting the landscape. When Leroy was farming in the 1960s, on the six-mile stretch of road where the farm is located there were probably ten families farming full-time, all of them with kids, and all shopping in town. Now there are two farm operators left, including John—one of whom (not John) works at another full-time job—and maybe two other farmers who lease large tracts of land in the region. Two of the former farms have been purchased as recreational properties for hunting.

When farms grow ever bigger, fewer people are left in rural communities. If farm families leave, taking their purchasing power with them, towns without other draws, such as tourism or manufacturing (which has also collapsed in rural America over the past twenty years), cannot stay afloat. And because of the tie between off-farm income and on-farm expenses, fewer jobs mean more farms will go out of business, further diminishing the rural workforce and potentially putting even large farms out of business.

Which would be fine if the end of rural communities was what the nation decided it wanted. But this isn't a future most Americans want to see.

"I think part of the American soul would die if we didn't think that there are cowboys somewhere in Wyoming," consultant Poppy Davis told me in mid-November 2020. Davis, a food business and policy adviser, was referring to the second great story of American agriculture: one known as the agrarian myth. This is the tale of rugged individualism that lies at the heart of our American ethos of independence and democracy. It's the narrative of farmers and ranchers as the embodiment of the American dream, agrarians who work hard to put food on our tables, who care for the land, and who are their own bosses.

As Davis pointed out, it's also the story that rural America will be there for us when we want it—that in the event of a pandemic or if city life gets too busy and overwhelming, we can always pack up and head for a small town with wide-open spaces, neighborliness, and simplicity.

Like so much in American agriculture, this myth too is almost as old as our nation. Thomas Jefferson, a founding father, farmer, and slaveholder, believed that ownership of land was one of our most desired and fundamental rights (albeit only for White men), the farmer a "bastion of freedom and independence, the vanguard of American democracy."[13] Those who worked the soil were "the most valuable citizens," he wrote, the "most vigorous, the most independant, the most virtuous, & they are tied to their country & wedded to it's liberty & interests by the most lasting [bonds]."[14]

But it doesn't take much knowledge of American history to see how the agrarian myth is disconnected from our country's real past, a fact that was clear even to Jefferson, who knew full well that the life he described didn't actually exist. "The pursuits of agriculture [are] the surest road to affluence," Jefferson wrote in 1787, "and best preservative of morals."[15] Yet even as he wrote these words, he was incurring enormous debt; farming never did make him the money he needed to support his land, slaves, and personal spending.[16] In fact, Jefferson died so deeply indebted that his entire estate, including the family's land—the Monticello plantation—was sold alongside his 130 slaves to pay it off. His grandson Thomas Jefferson Randolph paid the last installment on the debt owed by his grandfather fifty years later.

Laura Ingalls Wilder also worked to reinforce the agrarian myth with stories about her childhood in her Little House books, stories of an idealized existence that had little to do with the struggles her family actually endured.[17] In a fascinating book about the Ingalls family and their life on the prairie, Caroline Fraser tells of giant snowstorms that trapped the family inside their shack for months, land stolen from the Dakota and Osage tribes, and the annual destruction of crops by grasshoppers. Above all, the real Ingalls family history was about chasing an elusive agrarian dream; Laura's father, Charles Ingalls, bought property around the Midwest and moved the family repeatedly, sold on deceptive tales of highly productive land fabricated by speculators and the US government. Ingalls Wilder's writings about her childhood "were not only fictionalized but brilliantly edited," wrote Fraser, "in a profound act of American myth-making and self-transformation."

Much later, the *Little House on the Prairie* television series went several steps further in romanticizing the life of the Ingalls family. The show turned tar paper shacks into two-story wooden cabins with windows and turned the hard work of Pa (often depicted shirtless, as Fraser points out) into a success story of the small-scale agrarian on the midwestern plains, while in reality he struggled and ultimately failed in his pursuits.[18] The airing of the show, alongside others including *The Waltons*, *Petticoat Junction*, and *Green Acres*, helped cement America's fondness for quaint agricultural communities (albeit with less educated inhabitants) while ignoring the struggles that plagued many towns in real life. The simplicity and pureness of rural living, coupled with the virtuousness of hard work, were synonymous with the American dream.

Today, the myth lives on in YouTube videos of homesteading, on Pinterest pages of "farmhouses" equipped with rolling "barn" doors, and in Super Bowl advertisements for trucks.[19] And as farms have grown bigger, it is not only the farm lifestyle that we have turned into fiction; the images we have of the relationship between farmers and their animals is also now far from reality.

Brett Mizelle, a professor of history and American studies at California State University, Long Beach, wrote a book called *Pig* that looks at the contradictions between fictional depictions of the animal and its reality in American agriculture. "At the annual Pork Industry Forum I attended in Anaheim in March 2007, they held an auction to raffle off things to make money for the group," Mizelle told me. "All [of the raffle items] were images of single pigs or small farms; it was all nostalgia. And it was so strange, sitting there with a cocktail with these industry guys and asking, 'How many pigs do you have?' And they said, 'Oh, I have 30,000.'" The industry itself is caught between the bigger is better narrative and the agrarian myth. Even for those in the business, Mizelle added, it is impossible to conceptualize the reality of hundreds of thousands of pigs being processed each day in the United States. Instead we rely on images of single, cute pigs like those in *Charlotte's Web* or *Babe*.

"As I say to my students, you can't imagine someone doing a quilt of a gestation crate or a watercolor painting of a confinement facility," Mizelle mused. "Even the people in the industry are trapped."

~

Of course, there is nothing wrong with a little romanticizing—without romance I never would have been persuaded to move to Iowa. And, as with the bigger is better myth, there are truths embedded within the agrarian tale. There is great beauty in country living, and there is no denying that the quiet of the farm is deeply peaceful, the wide-open spaces freeing. There are certainly worse dreams than to move out to the country to farm, more harmful aspirations than to grow food or to improve the environment on a slice of Earth.

But the legacy of the agrarian myth has also trapped us in a particular mindset about farming that limits our understanding. It fuels the self-exploitation of farmers in Professor Ryan Galt's CSA study, and it is at the core of the individualism and pride in Donald Trump's bid to midwestern farmers. It's the reason so few farmers seek help for depression and anxiety and why we don't talk to beginning farmers about the financial realities ahead of them. And the image of an idealized small-scale farmer living the good life or the idea that larger farms make too much in government subsidies helps justify low prices at the grocery store while the people who grew that food continue to struggle year in and year out.

The narratives also make the existing cycle perversely sustainable. Even when treated terribly and paid next to nothing, farmers keep on keeping on, their economic concerns seen as petty in comparison with the virtuous missions of feeding hungry children or stopping climate change. If you get into trouble farming—these myths remind us—you should rely on your own self-sufficiency and high morals. And even when farmers are about to lose the farm to foreclosure, many still refuse to get help, the ideal of the resilient, do-it-alone farmer a stronger vision than the reality of their own failing economic or mental health.

Ultimately, when farmers buy into either or both of these myths, they are left blaming themselves. If we had planted earlier, had goats in addition to cattle, grown flowers and not just veggies; if we had been more efficient, bigger, bought that John Deere combine, used more chemicals; if we had offered pizza night or marketed ourselves better—if we had just done something else—then we could have been successful. Lack of income, most of us think, is a failure of our own character; those who can't manage the farm properly, who can't borrow more to make it through to next year, well, those are the farmers who don't deserve to farm anyway.

~

It is time for new story lines—not just stories about our own personal and farm experiences, but new national narratives. It is time to rethink who farmers are and what their work means to the security of our country, to reexamine a farmer's relationship to the land and to our collective physical health. And it is essential that the new stories include an honest look at the challenges of farming so that we can better define the problems and find real solutions.

To this end, the rest of this book examines four overlapping, highly nonscientific areas in which I think challenging these myths and creating counternarratives is critical if we are to someday find real solutions to our agricultural woes. First is the most basic, primal level, although arguably the hardest to change: the personal. Instead of glorifying the self-sacrifice and independence needed in farming, I will readily admit that my first experiences as a farmer were in fact full of stress and anxiety. There is no way to make change in a food system when feeling overwhelmed every day; belittling myself for not knowing all the answers—especially during the chaos of 2020—while perhaps normal, was not healthy.

Next is the farm level, the little slice of the world that John and I can help shape. Instead of trying to outcompete our neighbors, I will discuss ways for our farms to bring people together. Through collaboration, there is more opportunity to diversify the farm, to grow a more interesting variety of food, and to nurture new farmers.

Third is the community level. The bigger is better myth has catalyzed the collapse of millions of farms and the small towns that depended on them. Recognizing the interdependence of farms and the ways we can all benefit one another is the first step in rewriting our narrative about agriculture. I will look at two models of successful businesses that focus on collaboration between farms instead of competition and explore ways to strengthen bonds among growers.

Finally, there is the national level. With so many government programs already out there for farmers, how can policy help farms produce a steady supply of food while becoming more economically stable, more resilient, and better stewards of the land?

These four lenses helped structure my thinking and guided my search for ideas that follow in the coming chapters. And although I discuss them here in ascending order from the personal to the national, they are in fact not listed in order of importance, the four themes sloppy and overlapping like hogs at a feeding trough, blending into one another at times and distinctly independent at others.

My hope is that the concepts presented here can help redefine the way we think about agriculture and can serve to ground us in our quest to repair the system. I envision them as sparks to be taken out into the world—to farms, classrooms, and rural communities around the country—to add to the flame for change that is already burning. There is not a single idea I discuss here that can be instituted in a day or even likely in a year, and many have components that are already underway on other farms. But each of us can plant new seeds to grow over the years to contribute to the greater whole.

Chapter 12

Self-Care Is Key

In our second year of farming, 2020, the cattle looked good and were fattening faster than we had anticipated. The pastures bounced back quickly after each one-day graze, a helpful situation when it failed to rain for almost all of a very hot August. The riparian areas (previously called "ditches" by John and his family) were healthier than they had been in decades; we had spent days cleaning trash out of the banks—old tires, metal, shoes—garbage that Leroy, his father, and his grandfather had dumped. (What else did you do with trash before landfills and garbage pick-up?) We had stopped grazing cattle in the creeks, their mucking about in the banks a thing of the past, and planted hordes of bur oaks, northern pecans, chestnuts, and cottonwoods each spring and fall.

It seemed our efforts were paying off. Yet every decision, big or small, was exhausting. I felt as if I were trapped in a choose-your-own-adventure book, constantly worried that one false move would lead to disaster. We ran from task to task, our to-do lists streaming like endless stock market tickers.

A typical summer's day started at dawn and lingered past dark. Each morning we checked the cows and moved them into a new pasture, grazing them bit by bit across the landscape. Then the next day's paddocks needed to be set up and the previous day's rolled up—John, our nephew Lucas, and I rolling out the strands of electric fence and circling back to put in the posts (which, like everything, took far longer than it seemed it should). The fencer had to be connected and tested, and the mobile solar water system drained, moved, and set up again.

Then there were the bigger projects to tackle: hay to be cut and bales to be moved, permanent fence to be built, gates to be bought and installed. And looming in the background were the more nebulous long-term tasks: creating an email list and figuring out marketing materials, making calls to book times at the meat locker (almost a year in advance), and networking with other grass-finishing ranchers to learn more about how to raise the tastiest beef possible.

Sitting down to figure out our goals or profit margins never quite gained do-now status; taking time for planning felt like trying to herd frogs on a starry summer's night. We may have needed to build a new fence, but what was really missing was time and energy to grow into successful owners of a *business*, not just to complete tasks like neurotic farmhands.

Of course, 2020 did not help. It was an entire year of daily discomfort, the everyday drama of a global pandemic, an overdue racial reckoning, the crumbling of our economic system, and an American president intent on churning up our biggest fears. Uneasiness was all around, growing slowly like another animal we needed to tend to but didn't know how, keeping us both up at night.

~

Then one morning in July, I woke up with yet another litany of all we had not accomplished running through my brain. Why haven't you dealt with the yearling with pink eye? Where will you sell the cattle next spring? Why are you not working on the marketing?

It was a typical list—long and overwhelming—but this time the to-dos were accompanied by a tight feeling in my throat and chest, my heart racing even when I sat still. I felt as if I had just seen a tiger in the bathroom, but I was looking at a blank computer screen. I wondered if it was time to visit a doctor or, more likely with our terrible health insurance, to run off to the mountains and forget about this stupid idea of farming.

It was panic, of course, or at least her close cousin anxiety, that had taken up residence in my gut and chest. My constant focus on all that needed to be set right was driving my body into a state of high alert, a dull ache at the pit of my stomach, my breath shallow and taxing.

In my mind, the root of our problem was "the business." As I explained it to my mom (the CPA), farming felt like a Venn diagram, one circle filled with time-consuming but doable physical tasks like moving fence, the other overlapping circle occupied by The Business—a multitude of ambiguous duties, including marketing, that neither John nor I knew how to do and so never got started on. Like most farmers, our interests lay in the "doing" circle: being out in the field, taking care of animals, even driving around in big equipment. The idea of sitting at a desk while cattle needed to be moved, hay needed to be baled, and fence needed to be built felt like an indulgence. So the sitting down didn't happen.

As a freelance journalist, I was familiar with the need to constantly market oneself, but words are nonperishable, invented out of thin air, and honestly the pitching and promoting of stories was the part of the process I dreaded most. Reporting on a story is also an exercise in getting laser focused: learning everything possible about a single topic in order to describe it in minute detail. Farming, on the other hand, is more like juggling water balloons—many different tasks need attention all at the same time, each one amorphous and a bit slippery.

When we lived in the San Francisco Bay Area, John worked as a butcher and meat sales guy for several shops, which gave him great experience in knowing the cuts people want and how to prepare them. But being your own boss is a different thing entirely, and while John is one of the hardest workers ever, communication and planning were not his strongest suits. He took a few classes on the business of farming before we left California, but the students never practiced with the spreadsheets and tax forms on their own farms, so the information drifted off like pesticide in the wind.

Yet even if our business skills had been top-notch, it's

likely things would still have felt out of control because in farming so much is, in fact, out of your control. Even in the best of times, everything from the weather to world politics impacts your bottom line, not to mention your daily schedule. This, I believe, is one of the main reasons so many farmers have turned to genetic modification and company contracts, chemicals and government programs, "solutions" that give them a feeling of control over the elements and help create certainty in their lives. A contract is a guarantee about the future, even if it only guarantees how bleak that future will be.

But from twenty-five years of reporting on agriculture and from researching this book, I knew quick fixes were not the answer. There was nothing I could buy to ease the stress of farming, no scientific breakthrough that would mean great weather and long-term profitability. What really needed to be fixed, I realized, was me. I had to develop a firm personal foundation that would allow me to handle, and even enjoy, all that farming might bring my way.

Luckily for me, when the chest pains hit in the summer of 2020, I was not the only one out there struggling. Virtual sessions on everything from rocket science to basket weaving were readily available. And so I started with an email I found in my inbox one day offering a free time management course, part of a web series on resilience for women farmers.[1] I signed up and logged on.

"A lot of what I will talk about today is because I hit rock bottom on my farm and was like, 'Wait a minute, this is not how I thought life was supposed to be,'" Charlotte Smith, the instructor and owner of 3 Cow Marketing, started the webinar. Her words struck a chord immediately.

~

Smith's talk marked the first time in all my years of learning about sustainable agriculture that someone had spoken openly about how hard farming is, not just environmentally or physically but personally and emotionally. It is lonely work to patiently wait for the plants, animals, and land to heal in regenerative agriculture, for the slower-growing grass-finished cattle to come to weight, all the while worrying about where to process your animals and how to sell your products. Farming sustainably without the brand-name seeds and chemical regimens is tough work, but not just because weeding is more labor-intensive than spraying glyphosate. It is challenging because no one tells you step by step what to do; in farming, you are on your own—often both literally and figuratively.

But for beginning farmers the lack of honest discussion about the personal challenges of farming means that many of us experience paralyzing disappointment when we realize our big move to simplify our lives has resulted in chaos and overwhelming stress. We watched the Netflix documentaries and the YouTube how-to videos, and we read the books about the magic of farming. But when our lives don't match up to that agrarian dream, we come to feel like anomalies who just can't figure out how to crack the nut, how to make a living and stay sane.

Take the movie *The Biggest Little Farm*, for example. It's a beautifully shot film full of environmental revelations about how different parts of the farm can work together to create an ecological whole. But the movie covers for a grand total of zero minutes the enormous costs the farmers must have incurred in buying land outside of Los Angeles, building a state-of-the-art worm composting facility, and putting in hundreds of fully mature fruit trees with a fancy road winding up between them. How exactly did the farm survive financially when they did not sell anything for years after all the fruit was eaten by snails? The film never touched on the stress of juggling ten different farm enterprises all at once or the fact that someone with an outside job or family wealth had to be bankrolling the entire project. Sure, the pigs are dang cute, and who knew that ducks would eat so many snails? But why not mention the emotional and financial struggles the couple must have endured?

I want to stress that this is not just an academic point but an important paradigm shift. First, openly talking about the financial challenges associated with worm composting and planting trees in a movie to be shown at mainstream theaters and on Netflix could help change the conversation about farming on a cultural level. Every Joe Schmo could come to understand the economic costs of a farm, learning why some foods might be more expensive than others.

But it is also important for farmers to have role models online and on the big screen who are honest about what it takes to farm full-time. Many beginning farmers strive to emulate the sustainable systems we see, not knowing to take into account unforeseen complexities that could cost money we don't have. If we never hear about the expense of these operations or see the stress of making decisions when you have little experience, we won't know how to do it ourselves. When there are no road maps, no step-by-step programs to follow, each one of us strives to achieve that nirvana farm for ourselves. But few of us ever get there.

~

Charlotte Smith knew so much about the financial and emotional stress of farming—"overwhelm," she called it—because she lived it herself in her own dairy operation. So she taught a different kind of time management from what I had learned about before, one that had nothing to do with buying a good planner or making a better to-do list. It was about crafting your life into what *you* want it to be.

The main takeaway from her ideas, for me, was to plan out the week before I was in the throes of it. By taking the time to mindfully *choose* what I would do with my time beforehand, I freed myself from a lot of wasted energy (and anxiety) spent worrying about what to do each day. Not planning had made me feel as if I never got anything done, and it never allowed me to get to the projects I knew were important. Instead of being at the whim of mood or working on the next random thing that popped up needing attention, I decided ahead of time what was important and needed my attention. Once I committed to doing one thing at a time and doing it well—without beating myself up about all that I did *not* do each day—my anxiety eased almost immediately.

Buying our goats is a good example of how we put this theory into action. John and I wanted to get goats so that they would eat the invasive weeds (tasty food to them), one of those dream-farm land management practices. I was excited to meet and network with groups in the area who ate goat: the Eritrean and Nepalese, the Indian and Mexican communities, for example. John and I decided which kind of goats we wanted (Kikos) and from whom we would buy them; we called and reserved ten, to be picked up in the fall of 2020.

But as fall approached it dawned on me that if I was really going to raise goats well, it meant learning how to make sure they were happy and healthy and finding markets in which to sell them. This meant I would *not* have time for any other big project, including finishing a book or selling the beef we already had; there was just no way I would be able to remain focused while taking on another major task. We called Adam, the guy we were buying the goats from, and asked if we could delay their arrival until spring, when I would have time to properly focus on goats. We paid in advance (and extra, as he would need to feed and house the goats all winter), but I gained another six months to finish up my commitments before engaging in another. As soon as I admitted that I did not have the capacity to handle a new project (goats) and rescheduled it, the stress I felt about it melted away.

Taking Smith's workshop and online class, and listening to her podcast, gave me a good foundation. John and I also took a class with Holistic Management International, learned strategies to help us make better decisions on our farm, and ironed out our vision for the farm a little better. We renewed our membership with the Practical Farmers of Iowa, a farmer-led group that helped us feel immersed in more of a community. It was a place where we could turn to other farmers for advice and mentorship and made us know we were part of something bigger than ourselves.

~

Making these personal changes, learning to take better care of myself and to organize my time more efficiently, was a revealing process. I could see how I created my own anxiety through self-doubt and indecision. I also discovered that I had shrunk my attention span by flip-flopping between tasks, in effect making everything I did take a lot more time than needed. And I learned again that if I exercised every morning, my brain was much clearer and my mood uplifted.

Of course, not everything worked out perfectly, and some skills were (and are) slow to become habits. Setting a tight time frame for every single task, as Smith does, didn't work out all that well for me. It was too hard to transition quickly from one scheduled item to the next; I didn't have the time management muscles to keep on track and then felt guilty about not being more committed to my own commitments. To adjust, I scheduled bigger blocks of time into my day instead of hour-by-hour tasks. And I found it also worked well just to create routine—to spend each morning the same way, to write for the same hours, to help at the farm at certain times and days of the week. That way I still eliminated too much choice and indecision in my day but didn't have to wrestle myself into tiny blocks of time.

But I realized too that part of my battle was with the stereotypes I had in my brain about who I was *supposed* to be as a farmer—and even more so, a rookie woman farmer. I should be content just to be on the land, I had thought, awed every day by my great privilege of being alive, as those homesteaders on YouTube seem to be. I should be rugged and self-reliant, a strong woman who can keep up with the boys. I had thought John and I should have had an innate wisdom about things we didn't know, I realized, and I had believed that reliance on Mother Nature would ensure the system would take care of itself. In other words, because I hadn't really heard anyone before Smith talk about how psychologically exhausting my new career choice was, I thought it was just my own weakness.

~

So I started to do a whole lot less. I allowed myself to rest. And we began to accomplish more each day because instead of fixing a broken widget that presented itself out of the blue, we worked on the blog post we had said we would finish. When it seemed like a good idea to roll out tomorrow's fence but the calendar said to go home and make phone calls, we actually did. I set boundaries around when I would be available for John's projects (he had to get his act together to ask me in advance to help instead of assuming I could drop everything whenever he needed me). And happily, the new schedule ended up giving me more time to read books and play the guitar, activities I rarely did before because I was so unfocused and all I did was work.

But the most important outcome was the reminder that things take time. There had been so much change in my life in the past few years that I had forgotten that most change comes slowly and might take years to bear fruit, just like the trees in our orchard. We weren't going to be the Biggest Little Farm in a year or even likely six; we didn't have the money to pump in, for one thing. But slowly creating our own vision for the farm and mapping out how to get there step by step—proceeding slowly and methodically—was a much more sane model than running from task to task all day, every day. It was true that some important tasks wouldn't get accomplished. But I knew that doing less each day would, in the end, lead to doing more—for the farm and for myself.

Chapter 13

Co-farming and Community

OUR AD COST $43 AND RAN ONCE A WEEK for a month in the *Albia Union-Republican*. As an added bonus, it also appeared in the weekly shopper, amid the cheap food flyers and ninetieth birthday announcements. It was short—an inch and a half square—and, we thought, to the point.

"Looking for alternative farming opportunities? In search of farmers to raise healthy food with us. Esp. bees, chickens, sheep + veggies."

We decided to run the ad for several reasons. First, we were leasing a lot of land from John's dad for the number of cows we had, but we didn't want more cattle. Cattle prices had already been low for years, and the pandemic had made things worse for ranchers. COVID-19 outbreaks at meatpacking plants slowed down conventional processing, and when the processors didn't have the need for so many cows, ranchers instead turned to the local network of small meat lockers, the kind usually reserved for smaller farms and deer hunters. Almost overnight there were more cows than places to process them, and, as I mentioned earlier, we had to book our processing times almost a year out.

The number of cows, calves, and yearlings we had was also ideal for our first years as ranchers. It was a manageable herd, about seventy animals total, from which we could learn a lot about rotational grazing and how to market beef directly to consumers, hopefully without killing ourselves in the process. Our acreage also gave us a lot of room to rotate the cattle from pasture to pasture and to bale plenty of hay—benefits that would come in very handy if we had a severe drought of the kind that Leroy frequently reminded us was due to arrive sometime in the not-so-distant future.

But the 530 acres could support oh, so much more than just our seventy head of cattle. Much of the land we leased from John's dad held forests or riparian zones (the areas around creeks and ponds, full of trees and wildlife), good for grazing sheep or goats, for picking wild mushrooms, or for keeping bees. There was a lot of open flatland too that could be used for growing vegetables and plenty of room for chickens or ducks and flowers.

Diversifying the farm was key to maximizing the true efficiency of the land; this we knew. A multitude of plants and animals could easily be cultivated at the same time, often in the very same places. Cattle and sheep could be raised together on the same pasture in a "flerd," for example. Goats eat many of the weeds cattle won't consume while the cattle chomp down the grass where parasites that could potentially harm the goats live. Poultry could follow a few days after the flerd too, eating larvae out of the poop and keeping the fly population down for everyone. And, if you wanted to get technical about it, this kind of diversity could also yield a more nutrient-rich acre than a single crop of high-intensity corn ever could,[1] allowing us to feed the world in a very different—and far healthier—way.

This approach creates a complete, self-sustaining cycle, a diverse ecosystem that mimics nature, where total domination of a single crop—fertilized and pesticided—*never* occurs. It was also a well-established model of sustainable agriculture by the time John and I came to the farm, used for eons by Indigenous populations and popularized by Michael Pollan's *The Omnivore's Dilemma* back in 2006. In Pollan's book, farmer Joel Salatin[2] raises a large array of animals and vegetables dependent upon each other, all working together in symbiotic bliss—an example of a farm that has been held up as the ideal for twenty years now.

But it is also a system that has become part of the agrarian myth, a romantic vision of flora and fauna doing most of the work for the farmer. Sustainable farms are today expected to have a million different ventures taking place all at once, the theory being that if farmers just nourish symbiotic relationships, the chickens and sheep and goats and bees will take care of the rest. The self-exploitation Professor Ryan Galt described in his study—the one in which very few farms made a profit when they were focused solely on the ecology of the farm—is rooted in this narrative, a story that prioritizes land stewardship over personal well-being, long-term financial stability, or even positive family relationships.

In the real world, John and I were acutely aware of the stress and labor (or the money to pay for labor) this system would require. How could we take care of chickens and goats while weeding vegetable beds, moving cattle every day, *and* marketing everything we grew? As it was, remodeling a house and writing a book while farming were way too much; adding more to the to-do list was a recipe for mental and physical exhaustion. Joel Salatin and Will Harris of White Oak Pastures—another well-known White farmer who inherited the land he farms—might be able to juggle twelve enterprises to improve the land and increase their profits. But they also have a lot of help—family members, interns—and cash and loans available to them to risk on their ventures. For us, putting ecological benefits and romantic visions of a fully diversified farm before our own financial stability and our sanity would be simply substituting one unsustainable system with another.

But what if we didn't have to do all the work necessary to diversify the farm? It felt as if the saner—and likely more fun—route would be to divide up the work and the responsibility for the animals and plants by finding people interested in farming alongside us. We could feed many birds with one seed, so to speak, if Whippoorwill Creek Farm was not just a sole venture but worked to repopulate and regenerate the land at the same time. So much more would be possible if we found other farmers interested in raising animals to eat the weeds, add nutrients to the soil, and pollinate the vegetables, people to join us not as employees but as owners of their own businesses. We wanted to create jobs—better yet, careers—in which people could work the land for *their own* businesses. And, if we figured out how to do it, perhaps the land would not be just ours in the end but could be used by the whole group in perpetuity.

~

As I mentioned earlier, nearly 98 percent of farms in the United States are considered family farms,[3] yet almost 40 percent of all farmland around the country is rented out by nonfarming landlords to those who operate the farms.[4] In Iowa, more than half of all farmland is farmed by renters, generating rental revenue of $3.7 billion[5] for the landowners. And because 80 percent of the land in Iowa is owned free and clear, when an older farmer passes away, there is little incentive for the family to do anything but continue to rent out the land and collect a yearly paycheck.

But this system creates very little incentive for the farmers renting to improve the soil or install infrastructure when they have only a yearlong lease. The vast majority of leases in Iowa are annual, even though most land has been rented to the same person for more than a decade.[6] So most farm operators opt to grow annual crops that need little to no long-term nurturing of the ground and can thrive on trucked-in fertilizer, never tackling longer-term issues such as erosion and carbon sequestration.

Plus, with this system in place, there is little way for those with no land to actually own some of their own.

Even though it is expected that 10 percent of the country's ninety-three million acres of farmland will change hands in the coming years, only an estimated 2 percent will be sold to those outside of the family.[7] If land continues to pass down generation after generation in the same—mostly White—hands, how are nonlandowning BIPOC (Black, Indigenous, and people of color) or non-land owning White farmers to get into the business?

Our idea was to shift this narrative, away from the current "what's mine is mine" attitude, to allow for others to become established on the land too. Instead of renting our land, new farmers could bring their own sheep or chickens to graze, for example, and pay only a small percentage of their earnings to help upkeep and improve the farm. Eventually, John and I imagined, the group of farmers would collectively decide on a form of landownership that would allow our children and grandchildren access to the land and the ability to farm if they so desired, while ensuring that the group's hard work to improve soil, plant trees, and establish businesses was not tossed aside for someone to grow row crops.

We also wanted the land to be available after our deaths to those who were working it with us—for as long as they and their families also wanted to farm—and to give them the ability to bring on more people as the whole thing progressed. Instead of becoming another family farm statistic, with the land reverting to nonfarming Hogeland family members to sell or rent out, we would form a different kind of group of people working together cooperatively. We termed the idea "co-farming," based on the concept of cohousing, in which all members own their little slice of the pie but also contribute to the communal aspects of the venture.

Of course, we were not the only—or the first—to consider how to create a community landownership structure. New cooperative ventures with a focus on empowering farmers and challenging the existing food system were popping up across the country, such as Chris Newman's Sylvanaqua Farms and the New Roots Cooperative Farm in Maine, run by a group of Somali refugees. A small movement was also developing for a concept called "agrihoods"—an interesting model in which nonfarmers own houses (usually very fancy ones) on active farms in order to make the cost of land more manageable.[8]

Yet there is also a deep legacy in the United States—often left out of history books and current conversations—of groups working together to better their circumstances while treating land and resources as a commons.[9] Long before a White person's foot touched North American soil, lands were often allocated by tribal groups to specific families or groups who managed the land, collectively hunting, harvesting, and dividing food among members.[10] Prior to the Spanish arrival in the Southwest as well, Mesoamerican cultures created and maintained networks of ditches to share the limited water resources needed to farm.[11] Even the early colonies used a communal land structure in which the inner areas around a settlement were collectively farmed while the outer regions were used for grazing livestock. Some scholars believe that it was not a privatizing of land by individuals that was the initial source of conflict between colonists and Native groups in early America but a clash of communal lands.[12]

More recently, in the 1960s, a powerful cooperative movement also took place in Southern Black communities, documented by Monica White in her book *Freedom Farmers: Agricultural Resistance and the Black Freedom Movement*. The Freedom Farm Cooperative, for example, was started by Fannie Lou Hamer in 1967 in Ruleville, Mississippi, after she was fired from her position as a cook and sharecropper on a local plantation for leading a group of African Americans to register to vote. With her job also went the home she and her husband rented from the plantation owner. "They kicked me off the plantation, [and] they set me free," White quoted Hamer as saying. "It's the best thing that could happen. Now I can work for my people."[13]

Hamer saw landownership, voter rights, and food security as intertwined: owning and working land was key to true self-sufficiency and political power. She began Freedom Farm Cooperative initially with money generated from speaking engagements and endorsements by various celebrities, such as Harry Belafonte, to feed the community. But the venture also served as a housing initiative and business incubator for local Black residents. Later, with funding by nonprofits across the country, Freedom Farm purchased 640 acres, and by 1972 the group had fed more than 1,600 families and developed innovative economic opportunities for members.[14] A Pig Bank was started, in which the cooperative loaned out pigs to individual families to raise and breed. Each family then deposited two piglets back into the bank once they were weaned. In four years, 865 families took part in the program, raising thousands of pounds of meat[15] and creating small hog businesses for all the participating families.

Sadly, after only a few years, when the recession of the 1970s hit, donations to the project dried up and the cooperative folded. In the end, the venture arguably grew too quickly; offering so many programs was taxing on the financial reserves of the group. In 1976 Freedom Farm sold the land to pay overdue state and federal taxes; the burden of owning land while surrounded by racist communities actively working against the cooperative—delaying loans, delivering seed late, and the like—became untenable.

The North Bolivar County Farm Cooperative (NBCFC) in Mississippi—also highlighted in White's book—approached landownership a bit differently. In the first year, members worked voluntarily in the evenings or on weekends without pay. Proceeds went back into the group to buy seed and to rent tractors for the following year, and the cooperative utilized experts to advise on everything from growing techniques to nutrition. The co-op held listening sessions to hear about products community members wanted to grow and formed a partnership with Black transplants to Northern states to grow Southern specialties for them.

But instead of buying land, the NBCFC leased and borrowed land from regional Black landowners and opened a cooperative store in each of the twelve subsections of the co-op to sell produce back to the community. This meant the group avoided large capital land costs, reinvesting in seeds and processing instead of spending all its money to pay off mortgages and debt. The co-op also invested time and energy to find markets for its products before deciding what to plant.

And on a bigger scale, New Communities—started in the late 1960s in Georgia by Shirley Sherrod and her husband, Charles—raised more than a million dollars to purchase almost six thousand acres of land, becoming one of the original land trust models in the country. The cooperative remained successful until 1985, when droughts brought destruction and the cooperative was denied emergency funding from the US Department of Agriculture during the farm crisis. "You'll get a loan here over my dead body," Sherrod reported the county supervisor told her in an NPR story about the cooperative.[16] But years later, after a long fight in the *Pigford v. Glickman* class action lawsuit,[17] the group received restitution, and in 2011 it purchased a 1,600-acre piece of ground that once was owned by one of the largest slaveholders in Georgia. Today the group runs a research farm, provides assistance to farmers, and offers cultural and educational events.

Today, a handful of nonprofit groups and state governments are also working to get those with limited access onto the land using conservation and community easements, land trusts, and the rematriation of land. The Agrarian Trust, for instance, advocates for the purchase of lands to be held in common by a local Agrarian Commons chapter and rented out to farmers or groups of farmers. Similarly, the Northeast Farmers of Color Land Trust is creating a hybrid model in which land will be purchased and owned collectively by the trust and leased to BIPOC farmers on long-term leases. The State of Rhode Island has piloted a program in which the state buys farmland, protects it with conservation easements, and resells it to farmers at a reduced cost. And LANDBACK is an Indigenous-led movement that encourages White owners to return land to local tribes around the world, putting more land back into Native control for long-term agriculture, forestry, or other projects.

In Iowa, the Sustainable Iowa Land Trust (SILT) assists landowners in protecting their land with a conservation easement. But landowners can also donate land while continuing to live on the property. When the existing owner passes away or retires, SILT finds someone to farm the land, allowing the new operator to buy the buildings on the farm with no down payment and no interest. When a farmer or the farmer's heirs want to leave farming, they sell the buildings back to SILT, and SILT will make the farm available again to a new farmer. Land protected only with conservation easements (not donated) remains privately owned, but it too becomes more affordable for future farmers because of the restrictions placed on the land; it cannot be sold to developers, for example.

~

All of these ideas inspired me, and I could envision our farm adopting parts of these models. One option for us could be to separate the land from the farm business, keeping the land in family ownership for the foreseeable future while it is farmed by a cooperative. If the farmers themselves were spared the cost of land and did not need to pay mortgage or rent, the entire endeavor would become more feasible, I thought. Over the longer term, I hoped to convince John's family to consider leaving the farm to someone who currently does not have access to land and wants to farm sustainably, whether that person is in the family or not. But since the land is not mine (and never will be), I could see this might be an uphill battle. In the meantime, I would keep looking for other people to farm with us and keep learning about trusts and easements, return of land to tribal control, and collective landownership, so as to be able to better inform family members when the time is right.

Like the cooperatives of the 1960s, there would be other clear benefits of working together in a group. We could grow our own food, saving us each hundreds—if not thousands—of dollars each year and improving our access to healthy products. Sharing in the farm upkeep would ease the burden on John and me to care, and pay, for everything from fencing to tractors while making it possible for our colleagues to access machinery. Perhaps in a few years, I dreamed, the group could afford to rent or buy a facility in town where we could all communally process products and sell directly to consumers. Maybe someday we could even have a little restaurant.

Above all, working with other people would provide some much-needed camaraderie and social and cultural support. It could provide new jobs for those in our rural community, while improving the farm's efficiency, without any one of us taking on more than we could handle. Each farmer could leave occasionally to take an actual vacation and still be confident the plants and animals would be cared for by someone who had a vested interest. In other words, the collective would become the antithesis of the mythical American farmer toughing it out alone (and usually losing). We would have people with whom to weather the storms and celebrate our successes.

Yet even as I articulated the idea, I could feel the naivete of my words. I was on the phone with Suzan Erem, founder and executive director of SILT, explaining my ideas for our farm when I realized I sounded like a college kid, full of idealism but little life experience. I read her my long run-on statement about creating a place where we could farm, and so could other people (perhaps even farmers of color), and we would charge a small fee for them to farm, and allow them to stay on the farm forever, and maybe we could leave the land to them, if the family agreed . . .

"I will warn you, young farmers want certainty," Erem gently nudged. They might be polite Iowans and not tell me the idea sounded wishy-washy, she continued, "but they would be thinking the whole time, 'What is going to happen when some fourth-generation Hogeland shows up and wants to farm?'" New farmers deserve the opportunity to build a long-term enterprise and a life on the farm, she told me honestly, not move to a place full of unknowns. Where would the farmers live? How would we share equipment or sell products together? And would these other people be paying for the upkeep of the Hogeland land without any long-term benefit or equity for themselves?

Unfortunately, she explained, there is a multitude of horrific stories out there about landowners who changed their minds or leases that ended after the leasing farmers made expensive improvements. People want assurance and security in making a decision that will impact their entire lives, she explained, particularly people of color. They need to be able to trust.

In all my dreaming about the utopia that could be, I had left out the most critical element: relationships. When I read about the Black cooperatives of the 1960s and other farm groups such as kibbutzim and hippie communes, I had skimmed right over the fact that collectives usually grow *out* of a community and are rooted in deep trust of one's fellow members. Why would anyone trust our vision without a relationship with us, without knowing who we were and that we would be true to our word?

~

I was also rightfully challenged when I talked about looking for BIPOC people to work with us on our farm. "Society is shifting, and it is great that people have this heightened awareness and accountability to advocate for BIPOC people," Celize Christy, the next generation coordinator at Practical Farmers of Iowa, told me. But, she reminded me, while a lot of people like me (a.k.a. White women) want to help, for example, by making land available for farmers of color, that is not the same thing as working to remedy the reasons why people are currently not on the land. "People forget that it is not only establishing space, but making sure that space remains, that it doesn't go anywhere."

It was another aha moment for me in this journey. Asking "How do we bring people of color to the farm?" was not all that helpful if there was in fact no structural support for anyone to succeed once they were there. We could offer land, but that wouldn't mean the disappearance of a discriminatory lending system, nor would it reverse the history of racism at the USDA. It also wouldn't make Lovilia, Iowa, the town five miles away, a mecca of cultural diversity.

"People are eager to take action, but sometimes they don't realize what taking action means," Christy told me. "They don't think about how to be committed to the change long-term."

To this end, there are many things to consider. For one, offering the use of our land without any housing makes it unlikely that anyone could take us up on the offer. Housing is tough to find in a rural community such as ours; a farmer of any race would have to jump through many hoops just to live nearby. Figuring out a microloan system for someone who wanted to work on the farm would also be important, even if the enterprise was comparatively low-cost, such as raising chickens. And even if Lovilia turned out to be the most racially friendly place in America, the fact is that any BIPOC farmer who wanted to live there would be the only person of color in town.[18]

I called Shaffer Ridgeway to get his perspective too. Originally from Alabama, Ridgeway is a USDA district conservationist in Waterloo and is himself one of the seventy-two Black farmers in the state of Iowa, according to the census. In addition to increasing funding for beginning farmers of color, Shaffer had many other innovative ideas to add to the mix. "There are so many opportunities in agriculture—not just farming but in all of agriculture—and I want to see more African Americans in my field. It has been a great career for me," Ridgeway said.

To him, increasing the number of farmers of color starts with reintroducing youths to agriculture. Most young people are only one or two generations removed from the farm, Ridgeway told me, but many have heard terrible stories about farming from their grandparents. Facilitating relationships between farmers and young people could help recast agriculture in a more positive light.

Supporting more urban farms in Iowa might also increase the number of Black farmers without requiring people to move to rural communities and find housing, he continued. And getting more Black and other non-White people (and women, I might add) onto boards of national organizations such as the American Soybean Association, the National Cattlemen's Beef Association, and the American Farm Bureau Federation (all of which currently have none), Ridgeway added, would help create role models for aspiring farmers of color.

"I have been at my job for seventeen years, and everyone knows me," Ridgeway summarized at the end of our conversation. "My position has allowed me to get behind the veil with most of these farmers—I can get into the shop and talk with them on a personal level, because of my position. And I believe that to whom much is given, much is required. The privilege I have been given with my job to build all of these relationships with farmers, I think I can use that to help facilitate what I want to see happen with more farmers of color."

I agree wholeheartedly with Shaffer Ridgeway—John and I too must see our privilege as a responsibility. And while we clearly don't know all of the answers and will inevitably make stupid mistakes along the way, it is part of that responsibility to figure out how to assist new farmers to get into the business and stay there.

~

As was perhaps to be expected, our ad for farmers in the local paper and other posts like it did not elicit much in the way of responses. We did hear from a few people—a guy who had rabbits and hoped we might sell them for him; a Ukrainian national interested in farming with us but not in incurring any expenses whatsoever, even for housing; a vegetable farmer who needed land but found some closer to where he lived. There was an aspiring farmer who lived close by but wanted to raise cattle conventionally (we weren't interested), and a chef in Los Angeles I found via LinkedIn who was tired of city life and pandemic-related closures and was considering moving to Iowa. But that felt like a long shot.

"It is like being a yenta," Suzan Erem of SILT told me, using the slang Yiddish term for "matchmaker" to compare finding the right farmer to finding one's true love. Waiting for the perfect match was something that might take time, especially since our farm isn't located in a sexy area like Sonoma County, California, or the Hudson River Valley in New York. The farm sits an hour's drive from Des Moines, a long way away from anything unless you like the great outdoors and tiny towns with a single restaurant, a Dollar Store, and a population of six hundred.

"It is a thin needle to thread," Erem added. "Many people don't want to just farm; they want a way of life in a particular area where the hard work has already been done" (in New York, for example). "But I know that once we put our mind to it, we can find them." Maybe it would be a "California kid who is disenchanted, is sick of trying to grow food where there are fires and the weather is changing."

There have to be people interested in bringing life back into small, ailing rural towns, like the chef from LinkedIn or perhaps a passionate urban grower in nearby Ottumwa or Des Moines. People looking for more affordable land and housing and a chance to get out of the suburban nightmare of strip malls. Entrepreneurs ready for the challenge of taking a cheap storefront in a random town and turning it into a bakery, or converting an empty warehouse into a shop to make small-batch jam or yogurt.

~

All this is to say that we are, I am, still learning. Instead of being the mythical self-reliant farmers, interested in the short-term gains of the market, we need to keep our plans fluid and stay open to what is really needed to create a long-long-term success story. And while John's and my plan to offer up land for other farmers may have been a bit naive—especially when it comes to BIPOC farmers—the ideas are not stupid, nor are they completely unrealistic. At age fifty-two, John and I may not be able to experience for ourselves all that our farm can be, but we can try to create a solid foundation for the farm to grow and morph far into the future. Like the oaks we plant on our land, systemic change and equity need to grow and root over time. I want our farm to be in it for the long haul.

Chapter 14

Sharing the Pie

NICK WALLACE IS A FELLOW FARMER AND rancher who was in the business for at least a decade before John and I arrived on the scene: one of the who's who of sustainable agriculture in Iowa. We were talking on the phone about the idea of Iowa ranchers working together to market and sell beef. I mentioned creating a cooperative when Wallace posed an interesting question.

"But what do farmers think when they hear the word 'co-op'?" Wallace asked me. "Do farmers think it's a good thing?"

John and I met Wallace at a Practical Farmers of Iowa gathering of meat producers, a group that came together to see if we could figure out better ways to sell our products. Many farmers in the room marketed directly to consumers via websites and farmers markets—mechanisms that sometimes generated higher per-pound sales but often left farmers exhausted. Others sold to big processors via the sale barn or through feedlot contracts—a much less tiring way to offload cattle but one that produced profit margins so small they barely covered the cost of raising animals, if they did at all.

It was there that Wallace presented his ideas for a meat company. The business would purchase animals raised without chemicals for above-market rates, he said. Then he would sell the products via the internet to customers in nearby cities, like Omaha or Chicago. Customers could choose their cuts, he continued, while artisanal butchers in small towns would add value to the products, bringing entrepreneurial spirit and jobs to rural communities. Perhaps the company would even offer home delivery in trucks full of products consumers could pick from, much like the "Schwan man,"[1] Wallace mused, or maybe there would be brick-and-mortar shops in addition to meat lockers to process the product.

We were all excited about Wallace's ideas. Raising animals without chemicals or genetically modified crops could become more attractive to farmers around the state if they could sell to a company like the one Wallace described. And less chemicals, corn, and soybeans (and perhaps more livestock grazing on pasture to add nitrogen to the cycle) would mean a lower environmental impact on the soil, water, and air in Iowa, making us a more sustainable industry overall.

Many of us were also more than happy to outsource our marketing and sales, as we had neither the time nor the knowledge to successfully sell our products directly to consumers. From setting up processing dates to knowing the food safety laws to creating online purchasing platforms, direct sales took a whole lot of labor. While it felt like a giant waste of precious time for all of us to individually seek out and nurture customers, none of us wanted to sell to another company that treated us as nameless, faceless producers, all raising the same commodity product.

Wallace's ideas were also inspired by deep frustration. Consolidation had created a buyer's market for livestock, leaving farmers unable to negotiate a fair price. With power in the hands of so few companies willing to buy their animals, farmers have virtually no ability to set prices based on their actual costs. It is a problem that evolved over time as companies took over more and more of the process of getting beef from the farm to the plate.

"In order to have a better life—in order to stay on the farm and not work quite as hard—farmers relinquished the sales and marketing of their products to companies," Wallace explained. "But then those companies became bigger, and turned into even bigger stockyards and processors that sold to big grocery stores. And in the process the farmer got shafted. Today we have very little control over the markets anymore."

Farmers' lack of power in the marketplace is, of course, exacerbated by the "bigger is better" myth, which pits neighbor against neighbor. If growth is the only way to survive financially, smaller operations are necessarily overtaken or edged out. And with fewer farmers, the countryside becomes even more sparsely populated, the schools empty, the shops go out of business.

This cutthroat competitiveness also convinces farmers that installing a chicken or hog confinement unit is their fundamental right, despite what the neighbors think.

Even though putting in a facility lowers your neighbors' (and your) land values and increases the likelihood of antibiotic-resistant disease for those around it, a farm's private property rights trump all other concerns.[2] A community is no longer a group that works together for a common good; it is a loosely collected group of individuals looking out for themselves.

~

Even if agribusinesses and extension agents constantly tell farmers that competition is the only way to get ahead, strong evidence exists to the contrary. Cooperation not only is possible but can in fact yield the greatest financial gain, according to the work of Nobel Prize–winning economist Elinor Ostrom. Through examples around the world, from self-managed forests in Nepal to fishing waters off the coasts of Maine and Indonesia, Ostrom's work challenges the prevailing theory of the "tragedy of the commons." Despite economists' predictions that individuals will destroy resources as they strive to outcompete one another, often groups work together to protect the environment and reap financial success as a result.

Game theory—the study of how humans strategize in making decisions—reinforces this finding.[3] In a game called the "prisoner's dilemma," researchers have proven that people who cooperate—even if they are geographically dispersed and can't talk to one another—fare better financially than those working on their own. In other words, if farmers work to set prices *together*, *all* farmers will benefit.

With this in mind, Wallace's ideas seemed to me well suited to creating a cooperative. Each farm could pay a membership fee, I suggested, giving the group a way to finance infrastructure such as processing facilities and distribution trucks. And as a cooperative, I continued, every farm would have an ownership stake in the business and could participate in decision making via voting. It would be *our* company instead of just Wallace's.

But more than that, coming together could give farmers power. As I described in the previous chapter, many models exist to guide farmers in forming cooperatives, groups created to address not only the economic issues in farming but also the cultural and social ones. Yet a single farm functioning in the midst of thousands of acres of corn and soybeans can go only so far in changing the world. It will take many farms coming together, to create real alternatives to the current dominant commodity structure.

~

I again turned to older, more established cooperative companies to learn what made them tick. Organic Valley is one such entity, a well-known and successful agricultural cooperative whose products are found in nearly every supermarket in the United States. The company today boasts yearly sales of more than a billion dollars and is controlled by its nearly two thousand rancher-members, who elect representatives and are guaranteed a set price for their products. Back in 1988, when the co-op began, many of the farmers had the same complaints we have today.

"In the mid to late 1980s, it was a mess. The markets

were really terrible," Joe Klein, the midwestern dairy pool manager for Organic Valley, told me. Klein was referring to the farm crisis of the 1980s, when—yet again—the supply of commodities far overwhelmed the demand for them, and farmers were forced to sell their goods for less than the cost of production. "The group of farmers that started this organization were fed up," Klein told me. "They were like, 'What can we do to keep farmers on the land?' That was their number one concern."

The key, the group decided, was to get farmers a stable and fair market price, Klein said. Originally calling themselves the Coulee Region Organic Produce Pool (to sell produce), the group focused on a niche market, offering hard-to-find organic products directly to consumers.[4] The higher price customers were willing to pay for organics would cover the farmers' cost of production plus a small profit and allowed the group to guarantee farmers a set market price.

But how to structure the business wasn't clear. Many of the farmers had been part of co-ops in the past, but not everyone had had the best experience. "Many in the group were like, 'Co-ops?!'" Klein recalled. "We all know that co-ops struggle."

Since the early 1800s, farms in the United States have come together to form co-ops to purchase seed, fertilizer, and other products in bulk, achieving an economy of scale not possible for each individual farm. Cooperatives have also been created for farms to market products collectively or to have more control over the supply chain, processing raw materials together to receive a higher price. In fact, cooperatives have been so ubiquitous in agriculture that in 2014, 134 agricultural cooperatives in the United States celebrated anniversaries of one hundred years or more.[5] The majority of those cooperatives were located in the Midwest—Minnesota (35), Illinois (20), and Iowa and Kansas (each with 13).[6]

Yet, as Wallace astutely pointed out, many farmers today can tell you negative stories about cooperatives, just as the Organic Valley farmers did in 1988—stories about belonging to a group that sold seed and pesticides to members at a loss or about how the co-op couldn't compete in the commodity market and went under. Others gripe that their co-op struggled with organizational issues and power dynamics or about the increasing costs that plagued the group.[7]

"As they grow, cooperatives can lose focus," Klein told me. "When you first start out, it's great—you are supporting farmers, giving them a better price." But as the cooperative grows, the business might start to lose money because it is trying to do too much or decided to hire expensive executives to run the show. "Dissension can easily slip in," Klein added.

Beyond these standard disagreements are more extreme allegations of cooperatives that have engaged in nefarious activities. Farmer-members, for instance, accused the Dairy Farmers of America of cutting payments to farmers for their milk in order to make and sell cheese more cheaply. They also claimed the cooperative returned the bulk of the profits to investors rather than to its farmer-members.

On a more mundane level, perhaps the biggest issue is that cooperatives are notoriously slow-moving, unable to recalibrate or iterate quickly. The virtue of co-ops, their wide membership and emphasis on democratic processes, can also be their downfall. Presented with new opportunities or challenges in the marketplace, they tend to move like molasses as everyone meets, discusses, debates. Instead of taking risks, the group can take too defensive an approach, which, in today's anything-can-happen-at-any-moment world, might feel like trying to steer a bulky freighter through uncharted waters when a sleek speedboat is needed.

Even in the 1980s, getting farmers to join yet another cooperative was difficult. "I can't stress enough," Klein warned me, "people over thirty years of age have heard from their parents and grandparents about how co-ops fail. Getting past the word 'cooperative' means defining and redefining what your co-op is and how different you all want it to be from the others."

From its earliest days, Organic Valley tried to differentiate itself, beginning with strong leadership. "You need a really good leader," Klein told me, "a person who is knowledgeable. But someone who is also well respected *and* respectful to all members." Organic Valley also made sure to limit its activities. Initially the group decided to focus solely on marketing and selling products for a fair market price and left the rest to evolve over time.

Today, the Organic Valley cooperative is alive and well, even though just a few years ago (before COVID-19) the group weathered hard financial times when it couldn't sell all the milk its members generated. Cheap organic milk is now plentiful in grocery stores, presenting tough competition for the cooperative. But the group "tinkered" and grew as a result, which, according to research, is essential for cooperatives that thrive long-term.[8]

Klein contends the biggest indication of Organic Valley's success has been its impact on individual farms and communities. Towns that house Organic Valley offices are doing far better, he told me, than neighboring communities, which often struggle with vacant shops and few jobs. The guarantee of fair prices has kept farms in business too. "We truly have kept farms running, and the local economies are some of the biggest winners," remarked Klein. "There is no denying that."

~

Of course, Organic Valley is only one example of what a cooperative can be. One of the most attractive things about cooperatives is that there are no hard-and-fast rules about them, and yet one of the most confusing and scary things about cooperatives is that there are no hard-and-fast rules about them. Cooperatives can be set up as limited liability companies, as partnerships, or as cooperative corporations; members can market products together or share expensive machinery, transport items less expensively or purchase seed and chemicals as a group. They can be guided by social missions or purely financial ones, and they can allocate profits on the basis of investment or divide them equally among members.[9] Cooperatives can even be used as a part of a larger business venture, with hybrid companies created to accomplish a group's goals.

~

Guy Singh-Watson's business is just such a hybrid. He began Riverford Organic Farmers in the United Kingdom with a three-acre plot of vegetables, a wheelbarrow, and about £6,000 (about $8,000) in capital in the 1980s. Today Riverford boasts £100 million in gross sales per year (the equivalent of more than $136 million) and is majority owned by its one thousand employees.[10] The week before I spoke with Singh-Watson, in the midst of the coronavirus pandemic, Riverford had delivered 89,000 food boxes to points around the United Kingdom and, according to him, could have delivered as many boxes as the company wanted to.

"There are assumptions people make about what it means to be human, that we are motivated only by money," Singh-Watson told me. "But we need *both* competition and cooperation. And I think there is far too much emphasis on competition."

Like the founders of Organic Valley, Singh-Watson began his business by bringing together a group of like-minded farmers who wanted to grow food organically. But it quickly became apparent that his "big ego" (his words) was not well suited to the slow nature of co-ops. "I set the co-op up and really dragged them kicking and screaming in the direction I wanted to go," Singh-Watson told me, "which is not a great way to do it. And indeed, about two years in, they voted me off board."

Initially furious about his expulsion, a few weeks later Singh-Watson felt "immensely relieved" and started the work of creating Riverford Organic Farmers to market and distribute the food grown by the cooperative. "With most organizations it takes that sort of pigheaded, determined entrepreneur to get it off the ground," Singh-Watson commented. "I had this sort of demonic need to prove myself. Looking back, it is hard to believe I was so driven."

The co-op was left to slowly chug along doing what it did best—grow food. And Singh-Watson's company grew more dynamic and took bigger risks, creating franchises to sell boxes in different areas of the United Kingdom, for example (a system they are now dismantling because they feel they have outgrown that part of the business). In the end, separating the two entities—the cooperative and the company—established a better working relationship between all parties and allowed the food box business to flourish.

"I am a big fan of *cooperation*, but I am not necessarily a fan of cooperatives," Singh-Watson told me. "I acknowledge the benefits of capitalism—like innovation and responsiveness, and the ability to handle risk, for example. But all of our trading relationships, be they with farmers, the customers we are delivering to, or our staff and co-owners, they tend to be long-term, trusting relationships. A business can be built on good human relationships and trust and be mutually beneficial."

~

Back at the Practical Farmers of Iowa gathering, neither Nick Wallace nor I nor the other ranchers in the room were clear about how we might work together. From our perspective as farmers, there were a lot of questions about how selling our beef to Wallace would be any different from selling to other companies that bought our animals. Would we feel committed to selling to him even in the inevitable tough years when he wouldn't be able to offer us the best price? And from Wallace's end, the amount of capital needed to build the infrastructure his company would need—such as slaughterhouses, delivery trucks, and warehouses—was a paralyzing part of the puzzle. He also didn't trust that farmers would want to be part of a cooperative or that creating one would allow him to run the business in the way he wanted to.

And yet it is clear that there is an opportunity to create something larger than what a single small farm can offer customers but far smaller than the current system in which ranches are exploited by huge processors like JBS or the National Beef Packing Company. The fact that consumers voluntarily turned to smaller-scale, local producers for food when the coronavirus hit could mean there will be an ongoing market for our products. Plus, the level of frustration felt by ranchers is not going away anytime soon, even if prices rise for a time, because the current system gives farmers no power. Smaller processors also struggle with the current system because of how irregular business can be when dealing with smaller farms; working with Wallace's group could help guarantee a steady customer base.

As John and I got ready to bring our first beef to the local processor, we could see how much work lay ahead of us: all the phone calls and emails, the cut sheets and mathematical calculations. We wondered if all the work was going to be worth it, if the amount of time we would spend would really make us all that much more money in the end. We were excited to talk to people directly about our beef, about the way we take care of our cows and our commitment to improving the land. We even made a logo and had stickers and T-shirts printed. But we also hoped someday to work more collectively with other farms around Iowa, and perhaps with Wallace, in a venture to promote better-raised meat and make it more available to consumers. Together, we knew, we could have a much larger impact on the planet.

Chapter 15

People and Policy

STACY PRASSAS ARRIVED AT THE FARM IN A white Ford F-150 pickup, boots on her feet, baseball cap on her head, and giant earrings hanging down to her shoulders. A woman in her early forties raised with horses and dairy cattle, Stacy had worked for the regional Natural Resources Conservation Service (NRCS) office for seventeen years by the time she visited us at the farm. We engaged in the requisite Iowa chitchat about the weather while standing together around her truck, and then John and I carted Stacy over to the pasture in the four-wheeler to show her our problem.

The issue was our soil. Leroy had kept cows in the same pasture for several months every winter and spring to calve, and now, fifty years into the practice, the pasture was a muddy, manure-filled bog in the spring, a no-cow's-land. It was an area of the farm that kept me up at night worried about the cattle crossing it every time they wanted to get a drink of water or to eat some hay; we found one of our calves there, half dead, one cold, rainy day in March when he got stuck in the crook of a tree root in armpit-deep mud. (We brought him in and put him under blankets, and he survived.)

Stacy looked at the pockmarked ground destroyed by thousands of hooves stomping across it year after year. "The area is too compacted," she explained as she knelt down and picked up a piece of dirt (more like a glob of old cow shit) to show us that it stuck together like a slab of clay. The compaction led to standing water, she told us, and thus a muddy morass when it rained.

After talking about what to do with the area, John and I decided to seed it with Italian rye and field radishes in order to break up the dense organic matter with deep roots. Next year, or the year after, we would seed it with grasses and forbs and bring the cows back for a day at a time to graze. Above all, the area needed rest, time to heal from such a long stretch of intense use, to let the plants and animals and the microbes thrive again and create a living soil.

But more important to us than the advice we gleaned from Stacy was the relief we felt that she existed. Finally, we had found someone from the local government office who was on the same page, someone who spoke our language and meant the same thing as we did when she talked about regenerative agriculture or rotational grazing. Soon after Stacy's arrival, Nate Rahe came to lead the local NRCS office as our new district conservationist, a young man who also had an interest in restoring prairie and grass-finishing beef. We breathed a sigh of relief.

~

The ability of a farmer to get help at the US Department of Agriculture is as dependent upon the people who work there as it is on the policy they enact. It is true that the Farm Bill holds the key to almost all decisions about funding and programming.[1] Various stakeholders—from nonprofit sustainability groups to the corn and soybean grower associations to private companies such as Cargill and Tyson Foods and the behemoth American Farm Bureau Federation—come to the table to haggle over what should be included in the bill. In the end, the bill dictates almost everything in American agriculture for the next five years or so, doling out money for a variety of programs, including those that help provide food for people who need it, the decoupled commodity programs I described earlier, specialty crop programs (for fruits and vegetables), and even highly specific initiatives such as the Emergency Citrus Disease Research and Development Trust Fund.

And because the Farm Bill plays a central role in setting the direction of American agriculture, there is always a lot of discussion about how it can be used shape the nation's food supply. Ideas such as paying farmers to plant cover crops or to sequester carbon, for example, or mandating parity (guaranteeing farmers a fair price for their products) as a way to improve farmers' bottom line, become hot topics.

But on a county level, changes in the Farm Bill take time to trickle down. It was not until 2020 that our local office implemented changes in the 2018 Farm Bill, and even then, staffers knew only about programs relevant to them. Employees could tell you how to sign up for price loss coverage, for example, but not necessarily how it worked. And as I mentioned earlier, although John and I asked repeatedly about organics, we were told time and again that there were no organic or sustainability programs "per se." And yet there were—the Farm Service Agency (FSA) has offered an organic cost-share program since 2002.[2]

Which is to say that, although the wrangling that creates the Farm Bill matters a lot for the nation's food priorities, implementation is also critical. The lag time between idea and execution can be excruciatingly slow, if it happens in your county at all. Getting employees who deal with constant administrative changes to learn every last detail of a bill just so they can assist the one farmer a year who asks about organics . . . well, it just isn't likely. And finding interested and committed government agents to promote *new* programs to farmers who might not even have organics or cover crops on their radar—that is a very rare occurrence indeed.

The policy changes made in the bill are, without a doubt, critical. Yet I worry that when advocates rest all their hopes for reforming the food system on policy, they forget that these are government agencies we are talking about. Offices that are notoriously inefficient, clouded in layers of changing bureaucracy and administrative issues. This is not a critique of our local USDA office (they are extremely nice people) but a fact of government in general, no matter where it is. Has going to the DMV (Department of Motor Vehicles) office ever been a great experience for anyone, or paying a parking ticket in person? There is a reason so few people attend city council and school board meetings—government procedure is bureaucratic and often tedious; democracy is messy.

Take our recent experience at the NRCS, for example, even with Nate and Stacy there to help us. We were awarded an EQIP (Environmental Quality Incentives Program) grant to help with improvements to the farm, namely, permanent fencing and a watering system. We were approved for both in late 2018 and we bought the equipment needed for the watering system by the summer of 2020, but then we were told we had to wait for another level of approval. There was an error in the paperwork, we were told; two different computer programs were being used, and our application was stuck in the middle. We waited. And we waited. We called about it repeatedly. About eight months later we finally asked for the phone number of the office where the money was stuck and started to call there instead. Two weeks later the check arrived and we could finally proceed with actually putting in the watering system we badly needed.

~

Despite these hang-ups, Stacy and Nate felt like a godsend. Before they arrived, we would enter the USDA office and employees would bring out forms already filled out to help us apply for money, never discussing the bigger picture of conservation, sustainability, or even long-term financial viability. The focus was on the immediate future—maybe at best a plan for a three-year rotation of corn and soybeans—never on the actual long-term goals of government funding or how a payment might help lead farmers in a direction that is better for the environment or for the community.

But many farmers desperately need education and advice, not just money, especially those of us who are new to the trade. The funding John and I received for an organic transition plan is a good example. We spent months trying to figure out how organic certification worked. When would we need to start applying for certification, and to whom should we apply? How much would it cost, and did it make financial sense to certify our cattle as organic? I called around to certifying groups, which referred me to websites where I could find a multitude of forms and cost lists. But even after my research, I was at a loss as to whether certification was for us; the whole thing seemed foggy and complicated. We needed advice.

We heard about the Organic Transitions Program and asked about it at the USDA office. No one there knew anything about it, and so before we knew it, we were supplied with the forms to apply, although we never had a single discussion about what we were actually applying for. A few months later we received word that our application had been accepted, and we were told to hire someone from a specific list to do the plan, a situation that I assumed (incorrectly, it turned out) would result in our learning more about the transition process. The woman we contacted from the list visited the farm (she was lovely), asked questions about what had been done on the fields in the past and what we planned to do in the future, and a few months passed. Then a hundred-plus-page Conservation

Activity Plan (CAP) arrived on our doorstep, templates filled out, boxes checked, and maps included. We could certify the cropland.

In the end, we learned nothing more about how to certify our cattle or whether it made sense to do so. No one at the NRCS office or anywhere else ever sat down with us to talk about what such a plan entailed or what we would do with it once it was written. We were never asked if the plan was helpful or if it advanced our goals on the farm, and after spending $3,000 on the plan, we still didn't understand why we needed it. It was never anyone's fault that things played out this way—no one intended any harm or withheld information—in fact, everyone was helpful throughout the process.

Yet, the same process has played out multiple times: we look for advice but receive help filling out forms and applying for money. Each program is full of requirements and ranking systems to qualify, and agents spend enormous amounts of time analyzing the color or compaction of the soil, for example, when we are applying for a program to increase wildlife habitat or to stop erosion on a streambed. The programs rank farmers who grow commodities but want to make small changes such as putting in a cover crop higher than farms like ours, with a holistic plan for regenerating the grasses and wildlife.

Instead of making funding available to us to create an organic transition plan, it would have been far more cost-effective to provide an organic specialist to sit down with us and talk through what we wanted to accomplish and what we needed to learn before handing us the money to do it. Farmers often need one-on-one consulting with someone who can walk them through the process—be it a somewhat straightforward concept such as organic certification or much larger problems such as access to land and credit.

One of the most important ways in which policy could impact farming would be to allow USDA employees to do more listening and problem solving and less paper filing. Agents should be encouraged to think critically about how to solve issues, to help farmers understand how their choices will affect themselves and the land. Particularly in places plagued by environmental issues, such as Iowa, the USDA or the extension agent is often the only resource farmers have outside of industry representatives. Our problems were of the type the USDA could remedy with a comprehensive farmer helpline, funding for one-on-one mentoring programs with agricultural experts, and support for more farmer-to-farmer groups such as the Practical Farmers of Iowa, in which more experienced farmers teach newer ones—usually out in the field, where farmers learn best.

~

Today the federal government also gives money to nurture a new generation of farmers—funding that is projected to reach $50 million a year by 2023 (although that is peanuts when compared with the commodity and insurance programs)—via the Beginning Farmer and Rancher Development Program. Nonprofits apply for funding, often to make classes available to address the kinds of technical issues we encountered—teaching classes on organic certification, for example, or how to properly balance the farm's books.

But for all the problems John and I encountered trying to navigate government programs, it is worth repeating—we are people with privilege. The land we farm is owned free and clear by John's family, and generations of Hogelands have received support and loans from the USDA. We have access to Leroy's machinery, often bought with borrowed money, and the land is in pretty decent shape, all things considered. We are surrounded by other families (all White) who have also borrowed money regularly from local banks and are known by name at the USDA offices. John and I also got lucky and found housing just up the road from the farm that we could afford, an unusual circumstance in rural America.

These advantages have roots in events that occurred more than 150 years ago. John's great-great-grandfather James Ship Hogeland traveled west surveying for the railroads in 1851. He made a living working for the same railroad barons who sold land to people such as Charles Ingalls via false claims, and when he finished, James decided he loved the rolling hills of south central Iowa and bought land. The same year he arrived, the 1851 Exclusionary Law in Iowa not only barred freed Black Americans from buying and owning land but "excluded" them from even living in Iowa.[3] Any "Negro" entering the state after the law was passed had to leave within three days of arrival (the law was declared illegal twelve years later). The year 1851 also marked the signing of the last Indigenous treaty in Iowa, with the Santee Band of the Sioux[4] (although the Meskwaki remained in Iowa after they purchased their own land), which sanctioned the land grab that had already been going on for twenty years. Even when people of color have owned land in Iowa—as in in the town of Buxton—that wealth was rarely passed on from generation to generation, the result of a mélange of discriminatory practices stacked against their long-term ownership of property.

Yet, even if one is privileged enough to inherit land, as we did, giant investments of time and money are needed to get a farm up and running; John and I have now put in more than $70,000, not including the cost of our housing. New farmers too must overcome deep and long-lasting stumbling blocks in addition to the structural issues of discrimination and land loss: enormous student debt and lack of rural housing, child care, and affordable health care.

Very few beginning farmer and rancher programs supported by the USDA address any of these broader issues, in part because huge systemic challenges are hard to fix. Postdoctoral researcher Adam Calo put it this way:

> The next time a prominent policy maker or food systems reform advocate launches into a call for new farmers, it is worth asking, "What happened to the old ones?" In other words, in order to "create" new farmers who will not quickly vanish or merely meet elite demand for organic foods, the forces that provoke loss of dignified and durable farming livelihoods must be identified and addressed. Encouraging

> new entrant farmers into [the current] dynamic makes no sense at all—it is akin to sending lemmings over a cliff.[5]

Calo thinks that much of the advocacy on behalf of beginning farmers is flawed, particularly when compared with efforts to address other social ills. While beginning farmer programs focus on the technical aspects of agriculture, "housing advocates would never ask people looking for housing to come to a webinar to learn how to fix their own ceiling," he told me via Zoom. Instead, housing activists make demands of the government for more housing or to change eviction laws. These are the types of asks, Calo pointed out, that agricultural advocates largely ignore when it comes to beginning farmers, the group who most consistently come up against all the system's failings.

To Calo's point, more funding and coordination among organizations could help farmers navigate tricky issues such as finding housing and credit. Incentives could be offered for landowners to sign long-term leases of five or more years to give farmer-tenants more security (Iowa allows up to twenty-year leases). Student debt forgiveness programs could be made available to beginning farmers—an idea that was introduced to Congress as the Young Farmer Success Act in 2019 but never voted on—and sustainable farming grants (not loans) could be awarded to offset the costs of converting row crops to more sustainable practices. In other words, instead of focusing solely on technical training, each program offering a hodgepodge of lessons about farming and accounting, the federal government could strategically fund initiatives to address more complex issues—the ones that make it hardest for new farmers to succeed.

A recent report, analyzing existing beginning farmer programs also found that the most successful programs had similar elements: they focused on one-on-one mentoring, allowed farmers to dictate the curriculum they wanted to cover, and created lasting networks for farmers to tap into long after the program ended. Other projects that received USDA funding lacked these elements, and most didn't collect data about the long-term usefulness of the information they presented. Interesting to me as a former educator, one of the key recommendations of the report was for programs also to implement general educational teaching methods—many projects present information without any regard for how people actually learn. Even these simple program fixes would help farmers more easily find answers to their questions and retain important knowledge for the future.[6]

~

Yet new farmers aren't the only ones in need of assistance. Many old hands find themselves reluctantly continuing to cultivate and sell cheap corn, soy, or wheat, even when they know the market is inundated. And, as agriculture writer Alan Guebert put it, "Everyone—especially those on Capitol Hill—knows the Donald Trump-Sonny Perdue Gravy Train, [which] has delivered an astonishing $130 billion in federal subsidies in just four years to American farmers and ranchers, is out of steam."[7] So what are farmers to do? While some may willingly ride out the inevitable tariff wars, the changing climate, and the loss of topsoil and fertility, others might gladly get off the treadmill with a little support and incentive. They need an escape from the boom-and-bust game of commodity agriculture.

The concept of a "just transition" has been applied to mining and other extractive industries in which employees need to make a change. The idea is that individuals who once provided a quasi-public service, such as coal miners, loggers, and, perhaps today, corn growers, should not bear the burden of changing careers alone. Agriculture has had a huge impact on the environment, much as coal mining has. And because food is something we all need and public policy is largely responsible for the fix farmers find themselves in today—promoting loans to buy expensive equipment that can do only very specific tasks, for example—taxpayers should take some responsibility for helping farmers adjust.

First used in Canada in 2019 to ease changes in the power sector, the just transition concept is now being discussed as a framework in agriculture. And although the discussion is only in its infancy, the principles of a just transition can help to ensure that commodity growers, in addition to feedlots and other environmentally questionable practices in agriculture, have the opportunity to shift their focus from getting bigger to becoming more sustainable. Instead of just prohibiting practices such as animal confinement, we could encourage the changes needed to benefit the environment and the community. Such a program could offer debt deferral, for example—the ability to put off loan payments for three to five years—as a farm moves away from commodity crops.

"A huge reason my dad never got out of growing corn or soybeans was that he couldn't afford to skip a few years of paying off debt," John told me when I asked for his thoughts on how government programs could help farmers with a just transition. Growing corn or wheat means that you take out a loan every year to buy the expensive seed and chemicals, John explained, and then pay back the loan at the end of the growing season. Had Leroy wanted to raise grass-finished cattle or grow organic beans instead of conventional ones, he would have faced a gap of time without an income. Grass-finishing beef takes an extra year before a farm can sell the cattle, and shifting to organic crops means three years without chemicals before the farm can be certified. Under the current system, farmers can't afford to go without farm income for even one season; government support could change that.

National farm policy must also stop ignoring the problem of affordable health care. Farmers today depend on off-farm jobs for insurance; free or subsidized insurance could allow beginning and experienced farmers to instead put all their energy into farming and marketing their products. And because farmers are in effect public servants, supplying the nation's food and protecting vast stretches of land, it seems only right to afford them the same health benefits that other government employees, such as those in Congress, receive.

But farmers are self-reliant and oppose government involvement in their lives, you say—a national health system is not something they would support. Not true, data show. Despite the agrarian myth's assumptions about the independence of farmers and the assumptions that they don't want a public health-care system, a study found that three-quarters of farmers believe it is important for the USDA to guide discussions about health insurance for farmers.[8] A proposal for a farmer health-care program administered out of land grant universities was considered during the 2018 Farm Bill talks, but it did not make it into the bill.[9] Additionally, a system run similarly to the Veterans Association, housed at the USDA offices, a place farmers are already familiar with, could make subsidized health insurance available to those who wanted to enroll.

The state of Iowa, currently Republican controlled, already offers expanded Medicaid coverage under the Affordable Care Act (ACA), offering free health insurance—with a $6,500 deductible—for couples who make less than $65,000 per year, a benefit that John and I readily accepted in order to make farming possible. Without it, the cheapest plan available would have cost us more than $20,000 per year. The American Farm Bureau Federation also offers a "health plan" for members that covers preventive care, just as the ACA does, and costs less than most insurance available in the marketplace. But the Farm Bureau's plan (not to be confused with "insurance"—a distinction the state of Iowa made in allowing such plans to *not* meet ACA standards) also has huge deductibles and gives the company the right to deny coverage for preexisting conditions or to raise rates because of them. Given the rising age of farmers—the average is now fifty-seven and a half years[10]—accessibility to affordable and consistent health care—without the ability to deny coverage for preexisting conditions—is essential to the food security of our nation.

~

Farms, no doubt, need targeted agricultural programs, and many of the existing ones could be beefed up (pun intended) to be far more effective, particularly when it comes to assisting beginning farmers and helping farms transition to more sustainable practices. But farms would also benefit immeasurably from more vibrant towns; rural development is a key way for policy—*and* private industry—to help improve agriculture.

Interestingly, most of the federal government's programs for rural development come through the Farm Bill and are under the direction of the USDA. That means the USDA is currently responsible for tackling huge challenges in rural America, such as housing and a lack of internet service and small business support—programs essential to the revitalization of rural communities but with which the USDA has little expertise. There's no unified plan, no mission, no clear goals, and no real commitment by the federal government to America's smaller towns. The Congressional Research Service documented eighty-eight programs administered by sixteen different federal agencies, each with its own specific aim, such as increasing rural housing options, addressing rural poverty, or improving communities' access to medical services.

Under the direction of President Donald Trump, Secretary of Agriculture Sonny Perdue headed the Task Force on Agriculture and Rural Prosperity, which toured the country and outlined one hundred recommendations for improvements, categorized into five areas, "e-Connectivity, Quality of Life, Rural Workforce, Technological Innovation, and Economic Development."[11] Since the task force issued its final report, many new task forces have been created, including the Task Force for Reviewing the Connectivity and Technology Needs of Precision Agriculture in the United States.[12] While precision agriculture—in a nutshell, the use of technology to deliver the exact amount of pesticide, nutrients, or even water a plant needs—arguably has some importance in small towns, it is certainly not the only or even the most needed use of technology in rural America.

I can tell you without creating a task force that access to high-speed internet service is critical for bringing new and interesting opportunities to rural communities beyond precision agriculture. In 2016, less than 70 percent of those who lived in rural areas had access to broadband, as opposed to almost 98 percent of those in urban communities.[13] If the government followed Purdue's recommendation to provide robust broadband connectivity to every small town across America—perhaps through the rural cooperatives that were set up to provide electrification and are still alive and well today—it would galvanize a whole slew of new economic opportunities.

High-speed internet has already brought unique offerings to rural communities. The Fab Lab at the Center for Innovation and Entrepreneurship on the main campus of Independence Community College in Independence, Kansas, is one such project. Independence has a population of fewer than nine thousand, and when the lab opened in 2014, "we initially just thought it would be a cool place for people to come and make stuff," Jim Correll, director of the lab, told me. Now the lab is a fifteen-thousand-square-foot makerspace filled with laser engravers, plasma cutters, woodworking tools, AutoCAD, video equipment, and more. Entrepreneurship classes are offered, and Correll works closely with the town's chamber of commerce to provide business advice and services, including start-up loans.

"We had a vision in the beginning to change the way we think about the workforce," Correll continued. "So much of training has been, 'We are going to train you to do this one thing and then we expect you to do it for the rest of your life,'" he explained. Today, with technology changing all the time, Correll added, we just don't know which careers are going to disappear and which will be hot in five years. Instead, his goal is to equip people with diverse technical knowledge and help them become entrepreneurial thinkers, "because working in an environment like this makes your mind open to new things," Correll told me. Today the Fab Lab is credited with helping to revitalize Independence's downtown and supports more than one hundred businesses in the region.

With additional infrastructure, small towns could develop many other strategies for creating opportunity. Some smaller towns, the *Wall Street Journal* recently reported, are offering relocation incentives to remote workers to boost population bases and bring in expertise.[14] Tech companies focused on agricultural improvements, such as precision agriculture, could employ local farmers as consultants, bringing resources into the community and possibly creating products farmers really need. Instead of outsourcing, industries of all kinds could "onshore" customer service to small towns across the United States, with reliable broadband and local support. Value-added food products could be also be made closer to the farms that grow raw materials in rural communities, thereby funneling more of each food dollar back into the areas that grew them.

Rural revitalization is a chicken-and-egg phenomenon—if an area has quality jobs, more people want to live there. More people with more income means more opportunity for commerce in a town: opportunity for entrepreneurship and small businesses such as restaurants and markets, consulting firms and food processors. Restaurants serve more people, who consume more food, which in turn creates new opportunities for farmers to sell products directly to them. Small-scale processing of food into value-added products, such as jerky or bread, becomes possible when there is labor to hire and advisers who can help small businesses get up and running. This in turn creates more jobs, possibly filled by farmers or their family members looking for off-farm work, which, as I discussed earlier, keeps most farms afloat. And more smaller farms could mean more people, which stabilizes the population, brings in tax revenue for things such as libraries and recreation centers, and allows schools to fill. All of this makes for a higher quality of life and people who want to live in a town, even young people, who today continue to leave in droves. And so the cycle continues.

When Stacy Prassas came out to the farm, we were excited to learn what kinds of grasses to plant in the muddy pasture to restore our field. But we were also grateful for a feeling of community, to know that there were supportive people to call for advice and to help us problem solve, not just to give us a form to fill out. There was someone at the USDA office still ready to learn, who wasn't sick and tired of new rules and computer programs, who was excited to help create a network of farmers who can help one another.

Rural America needs more Stacys, and the way to attract them, or bring them back, or keep them there, is not just a question of federal policy. Private industry too needs to see rural America for the investment opportunity it is—places with unmet potential for valuable workforces, for new product ideas, and for a diversity of thought different from what you can get in the city. County policy could encourage these partnerships too—not by giving giant companies tax breaks to create low-paying jobs which extract profits out of town for shareholders far away—but by providing incentives to companies looking to diversify a town's options and improve a community's quality of life. Instead of another Dollar Store or Walmart, let's attract companies that can catalyze and enhance our technical, service, and entrepreneurial talent. Because in the end, it is the community and the individuals within it that make a town a place where people want to live.

Chapter 16

Meanwhile, Back at the Ranch

"BETH. JOHN. I WANT TO TALK TO YOU." It was Leroy calling out to us, yet again, in a serious tone from the back door of his house. We entered, the screen door slamming behind us in just that way it does, the sound often a prelude to a soliloquy about Leroy's past.

"Now, I want to talk to you both about 1934," he started, once we were seated at the kitchen table, John and I a bit relieved there wasn't something more urgent that was wrong. "In 1934," Leroy continued, "it was dry. And I don't mean just dry. There was a serious drought and people were starving."

It was a story I had heard many times, but I had never before thought to look up what a "serious drought" meant. The American Meteorological Society in October 1934 reported that hot conditions beginning in May, coupled with little winter precipitation, had led to a "state of ruin not before known to white men."[1] Des Moines endured twenty-five days of over 100 degree temperatures, and Ottumwa, about a thirty-minute drive to the southwest of us, hit a high that summer of 115 degrees.[2] And while the nine inches of rain that fell that year did not break records for the lowest amount—the summer of 1886 had seen only a little over four inches—the combination of a dry winter and spring and high heat made for a disaster.

Leroy continued about how they went without rain for so long that there was no grass at all for the cattle. All of two years old at the time, Leroy recalled how the men walked the cows down to the creek about a half mile away each day for water. "And they would cut down a tall tree every day, so the cattle could eat the leaves," Leroy told us. Feeding the cattle tree leaves, he reported, saved the herd and saved the family from financial disaster.

"I want you two to be thinking about it, because we are due for a drought like that again, a hundred-year event, the likes of which you have never seen," warned Leroy, listing the years since 1934 when there have been more minor but significant droughts, an occurrence that seems to average about every twenty years. "You have that old slurry tanker that could be rebuilt to haul large amounts of water, if you need it," Leroy told John. "And it is always a good idea to have extra hay on hand."

Leroy's advice about a major drought on the horizon was easy to dismiss. He always had a story to tell like the one about the trees, stories about his family during the Great Depression and of hard times of the past. There are plenty of ponds now on the farm, I thought while he was talking, and a rural water system we could always tap into if times got tough. And if things got bad enough, in this modern era we could just sell everything and drive out of Iowa.

But in my dismissal, I missed the point. Leroy has fully accepted that we are now the caretakers of this land. He has passed the torch to us, and with it, he is passing along critical information from one generation to the next. There *will* be a drought, and likely a flood and a windstorm too, especially now with such climate uncertainty. The information he offers can help us plan ahead for an uncertain future and reminds us that the comfort of our lives today isn't the way it always was or will forever be.

The stories of how the family dealt with extreme conditions are important for us to know. So are the other tales I chronicled in this book: about how Leroy's grandfather and others "raped" the land and about how they shipped animals to Chicago via the railroad. About how Leroy tried to grow the farm bigger—putting in a silo, joining MoCo 10—and made or lost money in the process, depending on the fickleness of the market and whether everyone else in the area was doing the same thing. Leroy has shared with me his rationale for adopting higher-tech seed and sprays (less work and higher yields) and explained the benefits of raising pigs indoors. John's mom, when she was alive, also weighed in about the family's history, lamenting that she and Leroy always seemed to be on a hamster wheel of debt that they couldn't get off.

These stories are key for understanding the food system today. So often, we forget the value of such knowledge, shared around kitchen tables, about mundane and huge events alike. These stories provide an antidote to the myths about farming because they reveal people's true motivations for doing what they do and allow us to humanize those we may write off as the minions of big business, uneducated or in it only for themselves. For every story Leroy told me of taking out a loan to grow bigger or buy more, there was an accompanying story about how the project ended, often in ruin. Without an understanding of the past—even if the truth challenges our beliefs or exposes the mistakes of our ancestors—there is no repairing of the present. If we don't acknowledge how things really went down, we become dazzled by romantic stories about a time that never was.

Living on a working farm is not about making your life simpler. It isn't only about putting your hands in the dirt (although you certainly can) and magically feeling more grounded, or getting up each morning to enjoy the sunrise as you milk a cow. It also shouldn't be all about self-sacrifice or endurance, independence or ruggedness. Unless you are raising food solely for your family, farming is a business. It is not a hobby, even if you don't make much money at it; it's hard work, often both enjoyable and very stressful. And every farm is embedded within an industry full of extremely complex problems—problems that can begin to be untangled only if we understand the history of how we got here.

~

So how have John and I fared in navigating our way through the American agricultural quagmire? Before we moved, I spent long nights awake, worrying that we would spend all our money in the elusive pursuit of breaking even. I doubted we had the business skills needed to ever make a profit, and I worried about how we would afford basic things such as health insurance. And after starting on this book and learning more about the data on agriculture, I was even more concerned that there was no way our farm could make a profit when so many others do not.

What I've learned over the past two years is that we can "make it," but with some important qualifications. We can afford to pay rent, but only if it is low and on family-owned land, land on which we can make improvements that will benefit us for years, generations, if we choose, to come. We can afford health insurance but only because we live in a state that subsidizes it for those of us not making much money at farming. John, it also turns out, is a very rare breed—a chatty rancher (and good writer) who is also a chef and a butcher—an extremely helpful mix when you are trying to sell beef. And we've learned much more about business techniques, stress management, and planning ahead—skills that *are* possible to learn while you are farming, as long as you can find what you need to know when you need to know it, via mentoring and classes.

In looking ahead to the rest of this year, our third—if all goes well (a big caveat in farming)—we will gross $55,000 by selling twenty-one head of cattle and hay and custom grazing other people's cattle for a fee. That is well below the $350,000 mark the US Department of Agriculture uses to demarcate farm businesses of the size that reportedly make a profit or, by a different perspective, make enough from government payments to not lose money every year. If no other big tragedy occurs, after paying our last $10,000 cattle payment and our $13,000 lease, we hope to net about $25,000. This is pretty pathetic for as much work as we will put in over the course of the year, for the environmental changes we are making, and for the quality of beef we are raising. But as you well know by now, John and I will be able to swing working for a pittance, unlike so many other beginning farmers, because of families' wealth: we have family land to lease, money in the bank, and little debt to our names.

Ten goats will arrive at Whippoorwill Creek Farm in a few weeks, and we will breed them next fall for the first time. The total number of goats and cows we will raise in the future is a good question—keeping goats involves a lot of fencing, and we don't yet know how the local coyotes, which can kill goats, will react to them. Goats are far easier to handle, however, than cattle are, particularly when they get sick or hurt. Plus, there is apparently a good market for them.

Outside of the farm, we see other opportunities available too. Lovilia, Iowa—the town five miles away from the farm—has a population of 623, a number that has not changed much since the 1950s, unlike so many other towns. But today it has literally three places of employment: the gas station Casey's, a newish Dollar Tree, and a locally owned bar and restaurant known for its "broasted chicken." When John was a kid there were also two grocery stores, a bank, a hardware store, three bars, a laundromat, a feedstore, a used car lot, and a barber shop—all of which, except for a tiny bank branch office, are now gone. Aside from the grocery store building that is now "city hall," few of the structures are even still standing.

Yet the town sits along a main artery of north–south traffic in Iowa (Highway 5), and according to the Iowa Department of Transportation, 3,360 cars pass through each day. The next-closest town to the south is ten miles away, and the supermarket there is a terrible cramped (and dirty-feeling) mess with a parking lot four times the size of the store. Lovilia is also just an hour's drive from Des Moines, a city that is considered one of the most underrated in the country for tech industry jobs, jobs that typically pay well and are filled by well-educated, often young, people who are frequently also passionate about food.

We also have the great luck of having neighbors who care deeply about the environment. One neighboring farm had a large herd of bison and is owned by a family more excited about prairie grasses and oak savanna than kids are about pizza. Another family is rumored to have nontraditional aspirations for their farm, though we have not met them yet. A Pheasants Forever sign sits at the end of our gravel road, denoting the group managing a chunk of land for the preservation and hunting of pheasants (which means there is prairie and forest, not more corn), and a section of Stephens State Forest—the largest state forest in Iowa, covering more than fifteen thousand acres—is nearby.

With only 1 percent of all of Iowa publicly owned, it feels as if there are opportunities to attract outdoorsy (possibly tech) people from the city to the rolling hills of south central Iowa for a bike ride on the rambling gravel roads or a hike in the state forest.

Of course, those same people could stop in and eat a delicious grass-finished burger and drink a beer. They could shop for "made in Iowa" local farm products, like the tomatoes a guy in town grows or the pickles we canned last summer, spending cash in a rural community that needs it. And although John and I both know it is likely a terrible idea to open a market and restaurant—most restaurants fail, and the business consumes your life (much as a farm does)—we can't stop talking about it.

There just so happens to be this sad little eatery in Lovilia, the kind that is sort of trying but serves "home-cooked" reheated-from-the-freezer food, we assume mostly bought from Sysco or some other such distributor and thrown into the deep fryer. During the coronavirus pandemic the building has mostly sat dormant, occasionally offering a greasy brunch, a car or two parked outside. But John and I can both picture what the place could be: the wooden interior space decorated with white lights hanging from the ceiling, picnic tables out back for barbecues and small gatherings. The side room that currently holds a few miscellaneous antiques could be a small market with produce from our farm and from gardens around town, and a place to sell our meats. Instead of paying a fee to sell products, other farmers could trade their goods for use at the restaurant or for their help working in the shop.

The restaurant would feature a burgers-only menu (veggie mushroom burgers and, of course, side dishes too), open Friday through Sunday, when people might also have the time to ride bikes or take a hike. There would be space for meetings or weddings, a commercial kitchen available to people who want to make jams or pies for their own businesses, and a big space where John could teach butchery classes. We would not only be the people who raise your beef; we could also be the ones to teach you about it—how it was cared for, how to cut it up, and how to prepare a little of it to go a long way. The building could also become a community center where people meet, teach one another, and maybe even watch a movie or enjoy a band concert on a hot summer night.

The economic difference in selling our beef as hamburgers versus selling it as quarters, halves, and wholes directly to consumers is startling. Selling our beef in quarters, for example, pencils out to $1,880 net for one cow after we deduct our expenses. Selling the same cow as one-third-pound burgers for $8 each, and subtracting rent and labor, would leave us with something more like $6,000 net for each cow. Of course, there is a lot more to factor in with a restaurant and shop—for one, good labor is not an easy thing to find. But paying for the overhead of an actual shop would also mean putting money into the local economy, generating a few jobs, and paying taxes locally.

~

Future dreams aside for a moment, by the start of year three, there were also a lot of accomplishments to celebrate on the farm. After a fair amount of anxiety, we sold our first round of heifers and steers without much struggle—fourteen wholesale to small distributors and six directly to consumers, mostly friends and family. We added to the herd (now thirty cows) and decided to calve in the fall instead of the spring so that cattle will be at the right stage of life for fattening up in the pastures during the summer. We also started custom grazing other people's cattle on our land over the winter. That way, we are sure to keep the fertility of the hay and the cow manure on our property instead of selling it all off to nurture someone else's soil.

We finished building a lot of fence, took other fencing down, and planted somewhere in the ballpark of two hundred trees. John concocted a solar watering unit that allowed us to build paddocks for the cattle more easily—we can bring water from ponds to the cattle instead of allowing the cattle to go into the water, a practice that eventually ruins your ponds. We bought a 40 acre plot of land adjacent to our almost completely remodeled farmhouse up the road to ensure we had property to do whatever we wanted to in the future without John's family involvement, including working with others and willing it to someone outside of the family. And we are trying to figure out how to make more housing available on the farm for those who might want to farm with us—BIPOC farmers or not—possibly converting old grain bins into small houses, a project that would be really cool. But again, it all just takes money.

On a more personal note, farming has taught me a remarkable number of things in a very short time. It has taught me that you can't always be there, do the right thing, or even plan ahead for disaster—shit really does just happen. And although I have dealt with adversity before,

I never had to do so on such a consistent basis; the predictability of suburban life shielded me from many of life's surprises. Now I know in my bones that there are all kinds of problems in this world we can't fix that are not worth my middle-of-the-night worrying. If it was 113 degrees in 1934—before humans made a complete mess of the atmosphere—then it certainly could be again this summer. We can plan for it, but worrying about it won't change a thing, and getting a handle on that anxiety makes life richer. It's a lesson I wish I'd learned long ago, that stressing about everything takes a toll on your body and soul and does little to help you prepare for hard times.

Which leads me to my second big lesson—that figuring out the best use of time and energy on a farm, and in the food system more broadly, is key. I have learned that taking on less is actually more—more sane, more realistic, more humane and kind to myself—and in the end it allows me to be more successful at each individual task. It is easy to waste a lot of time and energy as a farmer—or as a nonprofit or government entity, for that matter—running from task to task, doing things because there is a grant for them, never getting focused on what really should be your priority. But there is only so much time in a day, and with so many things to be improved in this world, all of us had better slow down, get aligned with our comrades, and figure out our priorities so we can accomplish the change we want to see.

To this end, John and I have become more active in our community. John ran for a seat on the board of our local Farm Service Agency, which makes local funding decisions, and although he didn't get enough votes, he can run again next time and win. He's also going to take over his dad's seat as a trustee—one of three people in each township who work to settle minor disputes among neighbors. And while it is arguably not the most important job to be the one responsible for settling fencing disputes or allocating the tiny budget for taking care of old cemeteries, it is a way to be part of the local community and develop leadership skills.

I joined the board of directors of the Iowa Farmers Union, and I am learning more about how the organization works and how it might help improve life for farmers. John and I are also active members of the Practical Farmers of Iowa. Through the group we have taken classes from and connected with other farmers, and we both see great potential for this model. I am also working to create or join a group in Iowa to improve the possibilities of success for *all* beginning farmers, focusing on the systemic challenges of coming into agriculture without land, wealth, housing, local social or cultural support, or even health insurance. This is where I want to spend my energy, and I hope to find a good fit for my skills.

~

But what about changing the food system?

How do we move away from a historically based, commodity-focused system that barely covers the cost of growing food to one that is more supportive of diverse small- and medium-scale farms? How do we ensure tasty and nourishing food, vibrant rural communities, and a range of farmer backgrounds?

I wish I could tie up this book with a bow, saying how cooperatives will fix everything, or that if farmers take better care of the land or their own mental and financial health it will all work out. But a whole messed-up system cannot be fixed with a single idea or even multiple ideas. The way food travels to our table—most often through a long, convoluted chain that yields the farmer little—has evolved over hundreds of years and will not be untangled without attention to the whole web of issues it has created.

Yet as I outlined in the previous chapters, there is much that we can do as individuals, as farms, as groups of farms, and as a nation to improve the system, all of which start with changing the stories we tell about farming. We need to move away from the romantic tales of farming to understand that while farmers feed people and take care of land, they also need to be able to take care of themselves and their families. Instead of idealizing the self-reliant, self-sacrificing farmer, toughing it out in the field alone and beating her competitors, farmers have to know that we can work together—perhaps even to *limit* our own output—for the benefit of the land and each other. We are not islands surrounded by hostile waters but are an interrelated ecology, a network woven out of our collective history. And that history should not be cast as an idealized "time before" when farmers grew healthy food, pulled themselves up by their bootstraps, and lived a simple, happy life—it is a far more complicated tale of colonization, discrimination, and commodification. We can move ahead only by creating a new future, not by replicating a nonexistent past.

Policy should support coordination among and within farms, the maximizing of nutrients and micronutrients above yield, and the ability of new farmers to get in and stay in the game. But farmers too need to take a deep look at who is representing them and what those groups are actively promoting. The American Farm Bureau Federation, for one, is supported by many farmers but is fully committed to the "bigger is better" myth—yet at what cost to its members? All farms—regardless of size—should question whether the current system, upheld by many farmer groups, actually benefits them in the long term and how better organizing could bring them more power.

And customers? Instead of journeying to the countryside to ogle a quaint farm, assuming its owner is living the perfect, simple life, see your local farmer for what she is: a hardworking businessperson connected to a web of people, plants, and animals. Ask about the community-supported agriculture box—is the farmer buying from other farms at a loss?—and if there is too much food in it, lend the farmer your support in scaling it back. Offer to trade your business skills—marketing, accounting, event planning—with your favorite farm to help create a community. Rather than asking why prices are so high, consider whether they cover the cost of production; encourage farmers to discuss the challenges of the farm, not just the successes.

Farming truly takes a village—especially if farmers are

growing perishable items such as vegetables or flowers, or if they are farmers historically not supported by the system. Farms simply *cannot* function alone in the landscape and be successful. And the output can't just go one way—out from the farm to the consumers. The village has to see the farm as part of its community, ensuring the land is well cared for and its farmers are mentally, socially, and culturally supported.

These solutions need to grow up from the grass roots—not come from top-down policies—from the farmers and want-to-be farmers, the nonprofits and the academics, the customers and the processors (and even the government employees) who are passionate about food. We have power in our numbers, and although the issues are complicated, we can each work on one small part—building more housing or making credit available, working to improve infrastructure, or making the land more accessible. Then we must continue sharing with each other our challenges and successes, iterate our plans, and keep on going.

To the credit of everyone out there already working hard on these issues, it's important to recognize that so much progress has already been made—we are not starting from scratch. We know a lot about what works and what does not work, as we can see with beginning farmer programs, for example. It is essential to keep analyzing our actions, tailoring them to farmers and communities—not to government programs—and working together. And we must recognize that change happens in farm time, in the long years it takes to grow a tree or pay off the mortgage on the land, all while weathering terrific storms and lulls in the market. Slow and steady progress will win this race on many fronts, but only if we are ready for the long haul.

~

It was January 2021 when John returned from a trip to Des Moines with the steaks. He had picked them up from a guy who bought two heifers and a steer from us, and we were eager to taste the results of our hard work. I could recall how John had carted the animals away few months prior, the truck with the livestock trailer behind it driving off into the horizon. I remember feeling sad that day, my hand in another being's imminent death a new and frightening thing in my life. After all that time worrying about the cattle's well-being, we were now carting them off to their demise.

But they had lived a good life, John and I knew, the very best life, in fact, that a beef cow could live. This "one bad day" would in fact be a very bad one, but it was really the only one in their short lifetimes. And whether you eat meat or not, the miracle of it cannot be denied—the fact that Mother Nature takes grass and water and air and turns it into high-quality protein and fat is an astounding feat. No fake meat comes close to replicating this ingenuity, and it never will.

Yet for all our hard work, we really had no idea how it would all turn out. Grass-finished beef can be lean and dry because of its much lower fat content; the reason the industry feeds corn and soybeans to cattle is not just that it quickly puts on the pounds; it also makes meat taste good.

So, our first bites into our very first steak—almost two years after we had taken over the farm—was a concerning moment.

Conversation stopped as John and I chewed, our taste buds on high alert. After a long pause, John gave the verdict. "It's really good," the butcher and chef said, and I nodded in agreement. All the hard work, the worry, the money, the learning—it had panned out, or at least it did this time. *Perhaps our bet had paid off.*

Notes

Chapter 1. The Simple Life?

1. Tyler Harris, "Report: Average Farm Debt Rises over $1.3 Million," Farm Progress, May 25, 2018, https://www.farmprogress.com/management/report-average-farm-debt-rises-over-13-million. The article refers to a report by Nebraska Farm Business, Inc. I contacted the company via email, and the executive director, Tina Barrett, responded that the number is now $1.4 million, and the survey was done with farms whose "average acreage is 1,380. Average gross income is about $1.25 million," which, she added, seemed average to her. This is a larger gross income than most American farms bring in, but it is more common for grain operations in the Midwest.
2. Please see the following for a discussion of the decision to capitalize White: https://www.chicagotribune.com/columns/eric-zorn/ct-column-capitalize-white-black-language-race-zorn-20200709-e42fag6ivbazdblizpopsp4p2a-story.html

Chapter 2. Land Rich, Cash Poor

1. US Department of Agriculture, National Agricultural Statistics Service, "2020 Cropland Value by State: Dollars per Acre and Percent Change from 2019," August 6, 2020, https://www.nass.usda.gov/Charts_and_Maps/Land_Values/crop_value_map.php.
2. Wendong Zhang, "2020 Farmland Value Survey," *Ag Decision Maker*, File C2-70, Iowa State University Extension and Outreach, updated December 2020, https://www.extension.iastate.edu/agdm/wholefarm/html/c2-70.html.
3. Twilight Greenaway, "Costco's 100 Million Chickens Will Change the Face of Nebraska," Civil Eats, December 11, 2018, https://civileats.com/2018/12/11/costcos-100-million-chickens-will-change-the-future-of-nebraska-farming/.
4. Listed for sale January 20, 2020, at https://www.tractorhouse.com/listings/farm-equipment/for-sale/164562215/2019-john-deere-s780.

5. Richard A. Levins, *Willard Cochrane and the American Family Farm* (Lincoln: University of Nebraska Press, 2000), p. 39.

Chapter 3. The Land of Corn and Cattle

1. *Merriam-Webster's Collegiate Dictionary*, 11th ed., defines "commodity" as follows: a product of agriculture or mining; a mass-produced unspecialized product; a good or service whose wide availability typically leads to smaller profit margins and diminishes the importance of factors (such as brand name) other than price. Excerpted from https://www.merriam-webster.com/dictionary/commodity, accessed May 2, 2020.
2. National Park Service, "Historic Jamestowne: A Short History of Jamestown," accessed March 2, 2021, https://www.nps.gov/jame/learn/historyculture/a-short-history-of-jamestown.htm.
3. Maureen Ogle, *In Meat We Trust: An Unexpected History of Carnivore America* (Boston: Houghton Mifflin Harcourt, 2013), p. 7.
4. Ogle, *In Meat We Trust*, p. 7.
5. Ogle, *In Meat We Trust*, p. 4.
6. According to the USDA, almost 144 pounds of meat (beef, pork, lamb, and veal) were available for each American in 2019, less than 3 pounds a week. Jeanine Bentley, "U.S. Per Capita Availability of Red Meat, Poultry, and Seafood on the Rise," *Amber Waves*, US Department of Agriculture, Economic Research Service, December 2, 2019, https://www.ers.usda.gov/amber-waves/2019/december/us-per-capita-availability-of-red-meat-poultry-and-seafood-on-the-rise/.
7. R. Douglas Hurt, *American Agriculture: A Brief History* (West Lafayette, IN: Purdue University Press, 2002).
8. Caroline Fraser, *Prairie Fires: The American Dreams of Laura Ingalls Wilder* (New York: Metropolitan Books, 2017), p. 64. Fraser here cites Richard White, *It's Your Misfortune and None of My Own* (Norman: University of Oklahoma Press, 1991).
9. The Beeches, "The Crops of 1860, and Their Influence upon Commerce and Industry," *New York Times*, August 4, 1860, accessed March 3, 2021, https://www.nytimes.com/1860/08/04/archives/the-crops-of-1860-and-their-influence-upon-commerce-and-industry.html.
10. Fraser, *Prairie Fires*, p. 97.
11. Fraser, *Prairie Fires*, p. 97.
12. Fraser, *Prairie Fires*, p. 98.

13. Allan Nation, *Grassfed to Finish: A Production Guide to Gourmet Grass-Finished Beef* (Ridgeland, MS: Green Park Press, 2005), p. 17.

Chapter 4. The Price of Sustainability

1. Wet bales can heat up to 130 degrees Fahrenheit, releasing a flammable gas that can ignite. Dan Folske, "Check Your Hay!," North Dakota State University Extension, September 30, 2019, https://www.ag.ndsu.edu/burkecountyextension/news/check-your-hay.
2. Gene Johnston, "Six Cover Crop Benefits That Pay," U.S. Farmers & Ranchers in Action, September 29, 2020, https://usfarmersandranchers.org/stories/economic-sustainability/six-cover-crop-benefits-that-pay/.
3. A 2015 study found that fertilizer and manure runoff costs the nation $157 billion per year in damage to both humans and the environment. Anne Schechinger, "Farm Nitrogen Pollution Damage Estimated at $157 Billion Yearly," Environmental Working Group, June 9, 2015, https://www.ewg.org/agmag/2015/06/farm-nitrogen-pollution-damage-estimated-157-billion-yearly.

Chapter 5. Privilege to Farm

1. Wendong Zhang, "Who Owns and Rents Iowa's Farmland?," *Ag Decision Maker*, File C2-78, Iowa State University Extension and Outreach, December 2015, accessed March 18, 2021, https://www.extension.iastate.edu/agdm/wholefarm/html/c2-78.html.
2. US Department of Agriculture, Economic Research Service, "Farming and Farm Income," accessed March 15, 2021, https://www.ers.usda.gov/data-products/ag-and-food-statistics-charting-the-essentials/farming-and-farm-income/.
3. Iowa Department of Cultural Affairs, State Historical Society of Iowa, "Sac and Fox Treaty, 1842," accessed March 18, 2021, https://iowaculture.gov/history/education/educator-resources/primary-source-sets/right-to-vote-suffrage-women-african/treaty-sac-and-fox-tribe-1842.
4. Bill Sherman, "Tracing the Treaties: How They Affected American Indians and Iowa," *Iowa History Journal* 7, no. 5 (September/October 2015), http://iowahistoryjournal.com/tracing-treaties-affected-american-indians-iowa/.
5. Sherman, "Tracing the Treaties."

6. Meskwaki Nation, Sac & Fox Tribe of the Mississippi in Iowa, "Meskwaki Nation—about Us," accessed January 15, 2021, https://www.meskwaki.org/about-us/history/.
7. Meskwaki Nation, "About Us."
8. Iowa Pathways, "African-American Residents Recall the Challenges of Leaving Buxton, Iowa in the 1920s," excerpt from "Searching for Buxton," produced by the Communication Research Institute of William Penn University, Iowa PBS, 2011, accessed January 15, 2021, https://www.iowapbs.org/iowapathways/artifact/african-american-residents-recall-challenges-leaving-buxton-iowa-1920s.
9. Rachelle Chase, *Lost Buxton* (Charleston, SC: Arcadia, 2017), p. 123.
10. US Department of Commerce, Bureau of the Census, *Fourteenth Census of the United States, Taken in the Year 1920*, vol. 5, *Agriculture: General Report and Analytical Tables*, chap. 4, "Farm Statistics by Color and Tenure of Farmer" (Washington, DC: Government Printing Office, February 8, 1922), p. 198, http://usda.mannlib.cornell.edu/usda/AgCensusImages/1920/Farm_Statistics_By_Color_and_Tenure.pdf.

Chapter 6. Money Matters

1. The USDA withdrew US standards for livestock and marketing claims (not including "organic") in 2016. This announcement can be found at https://www.federalregister.gov. Companies such as Topline Foods (https://toplinefoods.com) use their own definitions, as in "Grass fed beef simply means that the cattle were *allowed* to forage and graze for their own fresh food" (italics added). Grain can also be made available to cattle in this situation. Producers can opt for a grass-fed certification through groups such as the American Grassfed Association that allow only a 100 percent grass diet.
2. A cow is a female that has had offspring; a heifer is a female that has not yet had a calf.
3. Steers are castrated bulls.
4. Cora Wahl, Lora Liegel, and Clark F. Seavert, "Strawberry Economics: Comparing the Costs and Returns of Establishing and Producing Fresh and Processed Market June Bearing Strawberries in a Perennial Matted Row System to Day-Neutrals in a Perennial Hill, Plasticulture System in the Willamette Valley," AEB 0052, Oregon State University

Extension Service, October 2014, https://agsci.oregonstate.edu/sites/agscid7/files/oaeb/pdf/AEB0052.pdf.

5. Jessica E. Todd and Christine Whitt, "Farm Household Income for 2019 and 2020F—September 2020 Update," US Department of Agriculture, Economic Research Service, accessed October 20, 2020, https://www.ers.usda.gov/topics/farm-economy/farm-household-well-being/farm-household-income-forecast/.
6. US Department of Agriculture, "Market Facilitation Program," accessed March 18, 2021, https://www.farmers.gov/manage/mfp.
7. Quotes from Kautsky and others are cited in Ryan E. Galt, "The Moral Economy Is a Double-Edged Sword: Explaining Farmers' Earnings and Self-Exploitation in Community-Supported Agriculture," *Economic Geography* 89, no. 4 (October 2013): 341–365, https://doi.org/10.1111/ecge.12015.
8. Food justice is an area of activism focused on the equitable distribution and production of nutritious, culturally relevant foods.
9. Galt, "Moral Economy," p. 359.
10. Galt, "Moral Economy," pp. 359–360.
11. Galt, "Moral Economy," p. 361.
12. Mike Madison, *Fruitful Labor: The Ecology, Economy, and Practice of a Family Farm* (White River Junction, VT: Chelsea Green, 2016), p. 15.

Chapter 7. What about Subsidies?

1. The USDA's Farm Service Agency describes the Conservation Reserve Program as follows: "In exchange for a yearly rental payment, farmers enrolled in the program agree to remove environmentally sensitive land from agricultural production and plant species that will improve environmental health and quality. Contracts for land enrolled in CRP are 10–15 years in length. The long-term goal of the program is to re-establish valuable land cover to help improve water quality, prevent soil erosion, and reduce loss of wildlife habitat." US Department of Agriculture, Farm Service Agency, "About the Conservation Reserve Program (CRP)," accessed April 28, 2021, https://www.fsa.usda.gov/programs-and-services/conservation-programs/conservation-reserve-program/.
2. Called the Agriculture Improvement Act when it was adopted in De-

cember 2018, it was projected to cost US taxpayers $428 billion for the period between 2018 and 2023. Congressional Research Service, "What Is the Farm Bill?," updated September 26, 2019, accessed June 28, 2020, https://fas.org/sgp/crs/misc/RS22131.pdf.

3. "Covered commodities include barley, canola, large and small chickpeas, corn, crambe, flaxseed, grain sorghum, lentils, mustard seed, oats, peanuts, dry peas, rapeseed, long grain rice, medium and short grain rice, safflower seed, seed cotton, sesame, soybeans, sunflower seed and wheat." US Department of Agriculture, Farm Service Agency, "Enrollment Begins for Agriculture Risk Coverage and Price Loss Coverage Programs for 2021," press release, October 14, 2020, accessed March 8, 2021, https://www.fsa.usda.gov/news-room/news-releases/2020/enrollment-begins-for-agriculture-risk-coverage-and-price-loss-coverage-programs-for-2021.
4. US Department of Agriculture, Risk Management Agency, "Crop Year Government Cost of Federal Crop Insurance Program, 2008–2017," accessed May 6, 2021, https://legacy.rma.usda.gov/aboutrma/budget/17cygovcost.pdf.
5. Isabel Rosa, "Federal Crop Insurance: Program Overview for the 115th Congress," Congressional Research Service, R45193, updated May 10, 2018, accessed March 8, 2021, https://crsreports.congress.gov/product/pdf/download/R/R45193/R45193.pdf/.
6. EWG's Farm Subsidy Database, "Total Market Facilitation Program Payments in the United States Totaled $23.2 Billion from 2018–2020," accessed March 18, 2021, https://farm.ewg.org/progdetail.php?fips=00000&progcode=total_mfp&page=conc®ionname=theUnitedStates.
7. Congressional Research Service, "U.S. Trade Debates: Select Disputes and Actions," July 2019, updated March 2, 2021, https://fas.org/sgp/crs/row/IF10958.pdf.
8. Congressional Research Service, "U.S. Trade Debates." The funds were paid through the Commodity Credit Corporation, a government-owned federal corporation with borrowing powers of up to $30 billion.
9. EWG's Farm Subsidy Database, "Total Market Facilitation Program Payments."
10. There are also checkoff programs for commodity products such as

beef, peanuts, and dairy. Checkoff programs also do research on hybrids, pest management, and transport of the product.

11. Beef Checkoff, "About the Checkoff," accessed March 9, 2021, https://www.beefboard.org/checkoff/about-checkoff/.
12. This amount came from EWG's Farm Subsidy Database, accessed March 8, 2021, https://farm.ewg.org.
13. EWG's Farm Subsidy Database.
14. EWG's Farm Subsidy Database.
15. Boyd Kidwell, "Dream Big, Yield Bigger," *Progressive Farmer*, December 26, 2017, accessed January 10, 2021, https://www.dtnpf.com/agriculture/web/ag/news/business-inputs/article/2017/12/26/iowa-farmer-sliced-300-bushel-tiling.
16. US Department of Agriculture, Economic Research Service, "Assets, Debt, and Wealth: Farm Sector Equity (Wealth) Forecast to Remain Flat in 2021," updated February 5, 2021, accessed March 9, 2021, https://www.ers.usda.gov/topics/farm-economy/farm-sector-income-finances/assets-debt-and-wealth/.
17. US Department of Agriculture, National Agricultural Statistics Service, "Iowa Ag News—Farms and Land in Farms," February 20, 2020, http://www.nass.usda.gov/Statistics_by_State/Iowa/Publications/Other_Surveys/2020/IA-Farms-02-20.pdf.

Chapter 8. The Cattle Runaround

1. Many sale barns and auctions sell a variety of animals, including sheep, goats, and hogs. See Iowa Department of Agriculture and Land Stewardship, "Iowa Livestock Auctions," https://iowaagriculture.gov/agricultural-diversification-market-development-bureau/iowa-livestock-auctions.
2. The USDA tracks sales at seven or so of the largest sale barns in the state and makes the data available online. For example, see US Department of Agriculture, "Iowa Weekly Cattle Auction Summary," April 26, 2021, https://www.ams.usda.gov/mnreports/ams_2167.pdf.
3. US Department of Agriculture, Economic Research Service, "Cattle & Beef: Statistics & Information," accessed May 7, 2021, https://www.ers.usda.gov/topics/animal-products/cattle-beef/statistics-information.aspx.

4. As mentioned earlier, in Iowa those who raise cattle are often referred to as farmers, not ranchers.
5. Willard W. Cochrane, *The City Man's Guide to the Farm Problem* (Minneapolis: University of Minnesota Press, 1965), p. 5. According to Cochrane, in 1963, of the $66.4 billion in food sales, $21.4 billion went to farmers and $45 billion went to distribution and marketing.
6. Patrick Canning and Quinton Baker, "Food Dollar Series: Documentation," US Department of Agriculture, Economic Research Service, March 17, 2020, accessed January 11, 2021, https://www.ers.usda.gov/data-products/food-dollar-series/documentation.aspx#primaryfactor.
7. Joe Fassler, "A New Lawsuit Accuses the 'Big Four' Beef Packers of Conspiring to Fix Cattle Prices," The Counter, April 23, 2019, accessed October 7, 2020, https://thecounter.org/meatpacker-price-fixing-class-action-lawsuit-cattlemen-tyson-jbs-cargill-national-beef/.
8. Colin Brown, "#FairCattleMarkets Twitter Campaign Launches," Northern Ag Network, September 19, 2019, accessed May 8, 2020, https://www.northernag.net/faircattlemarkets-twitter-campaign-launches/.
9. Mackenzie Johnston, *Interview with Shane Kaczor, Co-owner of Bassett Livestock*, video, 9:42, Fair Cattle Markets, posted May 4, 2020, https://fair-cattle-markets.com/2020/05/04/interview-with-shane-kaczor-co-owner-of-bassett-livestock/.
10. Tom Polansek, "U.S. Senators Scrutinize Meat Packers' Big Profits during Pandemic," Reuters, March 30, 2020, accessed March 10, 2021, https://www.reuters.com/article/us-health-coronavirus-usa-meatpacking/u-s-senators-scrutinize-meat-packers-big-profits-during-pandemic-idUSKBN21H38M.
11. JBS USA, "About Our Company," accessed May 7, 2021, https://sustainability.jbssa.com/chapters/who-we-are/about-our-company/.
12. Sara Brown, "JBS USA Bets on Bacon with $20 Million Expansion to Ottumwa Pork Plant," February 11, 2019, accessed May 7, 2021, https://www.porkbusiness.com/news/industry/jbs-usa-bets-bacon-20-million-expansion-ottumwa-pork-plant.
13. KTVO News Desk, "JBS and Plumrose to Invest $3.1 Million in Ottumwa to Support Local Community," October 13, 2020, accessed May 7, 2021, https://ktvo.com/news/local/jbs-and-plumrose-to-invest-31-million-in-ottumwa-to-support-local-community.

14. US Census Bureau, "Quick Facts: Ottumwa City, Iowa," accessed January 11, 2021, https://www.census.gov/quickfacts/ottumwacityiowa.
15. Ana Mano, "Update 2—Brazil's JBS, World's Top Meat-Packer, Posts Better-than-Expected Results," Reuters, August 13, 2020, accessed January 21, 2021, https://www.reuters.com/article/jbs-results/update-2-brazils-jbs-worlds-top-meat-packer-posts-better-than-expected-results-idUSL1N2FF2I9.

Chapter 9. Keeping Up with the Joneses

1. Donnelle Eller, "Iowa's Hog Confinement Loopholes Causing a Stink," *Des Moines (IA) Register*, June 11, 2016, https://www.desmoinesregister.com/story/money/agriculture/2016/06/11/iowas-hog-confinement-loopholes-causing-stink/85264592/.
2. Eller, "Iowa's Hog Confinement Loopholes."
3. Miller calculated an $8,000 monthly mortgage for a $750,000 loan.
4. Cactus Family Farms, "How Much Can I Earn?," accessed January 11, 2021, https://www.cactusfeeders.com/cactusfamilyfarms.html.
5. Cactus Family Farms, "How Much Can I Earn?"
6. Daniel Anderson, "Does Hog Size Impact Manure Production?," National Hog Farmer, May 13, 2015, accessed May 7, 2021, https://www.nationalhogfarmer.com/environment/does-hog-size-impact-manure-production.
7. Iowa has free subsidized health care for families earning below $65,000 per year. This means that if the family does not pay for the health care, the government does, adding another layer of subsidies available for low-earning farmers.
8. Arin Greenwood, "Pigs Are Highly Social and Really Smart. So, Um, about Eating Them . . . ," Huffington Post, June 15, 2015, updated December 6, 2017, accessed January 11, 2021, https://www.huffpost.com/entry/are-pigs-intelligent_n_7585582.
9. Sharon Ann Murphy most recently wrote *Other People's Money: How Banking Worked in the Early American Republic* (Baltimore: Johns Hopkins University Press, 2017).
10. The volcano was on Mount Tambora in Sumbawa, Indonesia. "Dark Days for the World's Economy," part 2 of a talk by Andrew Browning, author of *The Panic of 1819: The First Great Depression*, video, 15:00,

YouTube, July 11, 2019, accessed March 15, 2020, https://www.youtube.com/watch?v=q1fqubvSVGE.

11. Digital History, "The Growth of Political Factionalism and Sectionalism," Digital History ID 3531, 2019, accessed January 12, 2021, https://www.digitalhistory.uh.edu/disp_textbook.cfm?smtid=2&psid=3531.
12. Brian Rukes and Andy Kraushaar, *Original John Deere Letter Series Tractors, 1923–1954* (St. Paul, MN: MBI, 2001), p. 47. John Deere almost went bankrupt during the Great Depression and was saved by a huge purchase of tractors by the Soviet Union.
13. Lee J. Alston, "Farm Foreclosures in the United States during the Interwar Period," *Journal of Economic History* 43, no. 4 (December 1983): 886, https://www.jstor.org/stable/2121054.
14. Betsy Freese, "SF Special: How Contract Feeding Changed the Hog Industry," *Successful Farming*, November 25, 2019, accessed May 11, 2020, https://www.agriculture.com/livestock/pork-powerhouses/sf-special-how-contract-feeding-changed-the-hog-industry.
15. Freese, "How Contract Feeding Changed the Hog Industry."
16. Wayne G. Broehl Jr., *Cargill: Going Global* (Hanover, NH: University Press of New England, 1998), p. 16.
17. Broehl, *Cargill: Going Global*, p. 19.
18. Richard A. Levins, *Willard Cochrane and the American Family Farm* (Lincoln: University of Nebraska Press, 2000), pp. 5–6.
19. Deere & Company, "Deere Announces Fourth-Quarter Net Income of $785 Million and $2.368 Billion for Year," press release, November 21, 2018, accessed January 12, 2021, https://www.deere.com/assets/pdfs/common/our-company/news/deere-4q18-news-release.pdf.
20. Peter Waldman and Lydia Mulvany, "Farmers Fight John Deere over Who Gets to Fix an $800,000 Tractor," Bloomberg Businessweek, March 5, 2020, https://www.bloomberg.com/news/features/2020-03-05/farmers-fight-john-deere-over-who-gets-to-fix-an-800-000-tractor.
21. Cactus Family Farms is the pork division of Cactus Feeders, a company that also specializes in cattle feedlots. Cactus sells 100 percent of its estimated 750,000 hogs per year to Tyson Foods. Betsy Freese, "Want to Contract Feed Pigs? Here's What You Need to Know," *Successful Farming*, January 23, 2018, accessed April 28, 2020, https://www.agriculture.com/livestock/hogs/want-to-contract-feed-pigs-heres-what-you-need-to-know.

22. Doug Bock Clark, "Why Is China Treating North Carolina Like the Developing World?," *Rolling Stone*, March 19, 2018, accessed May 7, 2021, https://www.rollingstone.com/politics/politics-news/why-is-china-treating-north-carolina-like-the-developing-world-122892/.
23. Betsy Freese, "Successful Farming Exclusive: Top 40 U.S. Pork Powerhouses 2020," *Successful Farming*, accessed May 7, 2021, https://www.agriculture.com/pdf/pork-powerhouses-2020.
24. Freese, "Top 40 U.S. Pork Powerhouses 2020."
25. Kurt Lawton, "Taking a Look Back at the 1980s Farm Crisis and Its Impacts," Farm Progress, August 22, 2016, accessed January 11, 2021, https://www.farmprogress.com/marketing/taking-look-back-1980s-farm-crisis-and-its-impacts.
26. American Farm Bureau Federation, Market Intel, "Farm Bankruptcies Rise Again," October 30, 2019, accessed January 11, 2021, https://www.fb.org/market-intel/farm-bankruptcies-rise-again.

Chapter 10. Everybody Does It—off the Farm

1. Carolyn Dimitri, Anne Effland, and Neilson Conklin, "The 20th Century Transformation of U.S. Agriculture and Farm Policy," US Department of Agriculture, Economic Research Service, EIB-3, June 2005, https://www.ers.usda.gov/publications/pub-details/?pubid=44198.
2. Jessica E. Todd and Christine Whitt, "Farm Household Income for 2019 and 2020F—September 2020 Update," US Department of Agriculture, Economic Research Service, accessed October 19, 2020, https://www.ers.usda.gov/topics/farm-economy/farm-household-well-being/farm-household-income-forecast/.
3. In an email with the USDA, I asked why the median on-farm income plus the median off-farm income does not equal the median total income. The response: "The three numbers are medians and since the median farm income might not occur in the same HH [household] as the median off-farm income or the median total income, one cannot sum the two to get the total."
4. Mitra Toossi, "A Century of Change: The U.S. Labor Force, 1950–2050," *Monthly Labor Review* (May 2002): 15–28, accessed October 9, 2020, https://www.bls.gov/opub/mlr/2002/05/art2full.pdf.
5. US Bureau of the Census, *United States Census of Agriculture: 1954*, vol.

2, *General Report: Statistics by Subjects*, chap. 3, "Farm Facilities, Farm Equipment" (Washington, DC: Government Printing Office, 1956), p. 180, http://lib-usda-05.serverfarm.cornell.edu/usda/AgCensusImages/1954/02/03/1954-02-03.pdf.

6. US Bureau of the Census, *United States Census of Agriculture: 1954*, vol. 2, *General Report: Statistics by Subjects*, chap. 5, "Size of Farm" (Washington, DC: Government Printing Office, 1956), p. 344, http://lib-usda-05.serverfarm.cornell.edu/usda/AgCensusImages/1954/02/05/1954-02-05.pdf.
7. US Department of Agriculture, Farm Production Economics Division, Economic Research Service, "Labor Used to Produce Field Crops: Estimates by States," Statistical Bulletin 346, May 1964, p. 5, https://ageconsearch.umn.edu/bitstream/153771/2/sb346.pdf.
8. Alejandro Plastina, "Estimated Costs of Crop Production in Iowa—2021," *Ag Decision Maker*, File A1-20, Iowa State University Extension and Outreach, revised January 2021, accessed May 11, 2021, https://www.extension.iastate.edu/agdm/crops/pdf/a1-20.pdf.
9. Sun Ling Wang and Eldon Ball, "Agricultural Productivity Growth in the United States: 1948–2011," *Amber Waves*, US Department of Agriculture, Economic Research Service, February 3, 2014, https://www.ers.usda.gov/amber-waves/2014/januaryfebruary/agricultural-productivity-growth-in-the-united-states-1948-2011/.
10. Dan Miller, "Deere Launches High-Capacity X9," *Progressive Farmer*, August 1, 2020, https://www.dtnpf.com/agriculture/web/ag/news/article/2020/08/01/new-equipment-deere-launches-high-x9.
11. US Census Bureau, "Men & Women, Money & Work," May 29, 2019, https://www.census.gov/library/visualizations/interactive/men-women-earnings-gap.html.
12. US Department of Labor, Bureau of Labor Statistics, "National Census of Fatal Occupational Injuries in 2018," news release, December 17, 2019, accessed November 10, 2020, https://www.bls.gov/news.release/archives/cfoi_12172019.pdf.
13. National Sustainable Agriculture Coalition, "USDA Invests $14 Million to Train the Next Generation of Farmers," blog post, November 7, 2019, accessed November 7, 2020, https://sustainableagriculture.net/blog/usda-invests-train-next-generation-farmers/.

Chapter 11. A New Narrative

1. US Department of Agriculture, National Agricultural Statistics Service, "2017 Census of Agriculture Highlights: Young Producers," October 2020, accessed November 1, 2020, https://www.nass.usda.gov/Publications/Highlights/2020/young-producers.pdf.
2. Drew DeSilver, "10 Facts about American Workers," Pew Research Center, August 29, 2019, accessed November 1, 2020, https://www.pewresearch.org/fact-tank/2019/08/29/facts-about-american-workers/.
3. "Survey: A Third of Iowa Farmland Owners over 75," *Iowa Farmer Today*, updated July 26, 2018, accessed March 19, 2021, https://www.agupdate.com/iowafarmertoday/news/state-and-regional/survey-a-third-of-iowa-farmland-owners-over-75/article_967146c4-8ad2-11e8-8490-57ad9e8081ac.html.
4. Linda Foreman, "Characteristics and Production Costs of U.S. Corn Farms, Including Organic, 2010," US Department of Agriculture, Economic Research Service, EIB-128, September 2014, introduction, https://www.ers.usda.gov/webdocs/publications/43883/49018_eib128.pdf?v=0.
5. Bill Spiegel, "How David Hula Grows 600-Bushel-Plus Corn," *Successful Farming*, January 6, 2020, https://www.agriculture.com/news/crops/how-david-hula-grows-600-bushel-plus-corn.
6. Estimates for the amount of labor to grow corn in 2021 are actually more than two hours per acre. Alejandro Plastina, "Estimated Costs of Crop Production in Iowa—2021," *Ag Decision Maker*, File A1-20, Iowa State University Extension and Outreach, revised January 2021, https://www.extension.iastate.edu/agdm/crops/pdf/a1-20.pdf.
7. James Risser and George Anthan, "Why They Love Earl Butz," *New York Times*, June 13, 1976, accessed November 27, 2020, https://www.nytimes.com/1976/06/13/archives/why-they-love-earl-butz-prosperous-farmers-see-him-as-the-greatest.html.
8. US Department of Agriculture, Economic Research Service, "ICYMI . . . The U.S. Is Not Capturing the Growth in Global Grain Trade," October 11, 2018, updated August 13, 2019, accessed November 27, 2020, https://www.ers.usda.gov/data-products/chart-gallery/gallery/chart-detail/?chartId=93665.
9. Christina Stella, "Should Ethanol Go Back to Business as Usual af-

ter COVID-19?," NET News and Harvest Public Media, November 24, 2020, accessed December 9, 2020, http://netnebraska.org/article/news/1243423/should-ethanol-go-back-business-usual-after-covid-19.

10. Shelby Myers, "Despite Its Many Uses, Ethanol Demand Remains Stifled," American Farm Bureau Federation, Market Intel, May 8, 2020, accessed November 27, 2020, https://www.fb.org/market-intel/despite-its-many-uses-ethanol-demand-remains-stifled.
11. Macrotrends, "Corn Prices—59 Year Historical Chart," accessed May 11, 2021, https://www.macrotrends.net/2532/corn-prices-historical-chart-data.
12. Iowa Corn Growers Association, "Biochemicals & Bioplastics," accessed December 9, 2020, https://www.iowacorn.org/corn-uses/new-uses.
13. R. Douglas Hurt, *American Agriculture: A Brief History* (West Lafayette, IN: Purdue University Press, 2002), p. 72.
14. In the original text, "bonds" was written as "bands." Excerpt of letter from Thomas Jefferson to John Jay, August 23, 1785, Jefferson Quotes and Family Letters, accessed December 9, 2020, http://tjrs.monticello.org/letter/69.
15. James Phillips, "American Agrarianism's Answers to the Nation's (In) Securities," *Connecticut Public Interest Law Journal* 9, no. 2 (2010): 347, accessed November 21, 2020, https://cpilj.law.uconn.edu/wp-content/uploads/sites/2515/2018/10/9.2-American-Agrarianisms-Answers-to-the-Nations-In-Securities-by-James-Phillips-.pdf.
16. Lucia C. Stanton, "Debt," Thomas Jefferson Encyclopedia at Monticello.org, 2019, accessed November 20, 2020, https://www.monticello.org/site/research-and-collections/debt#footnote8_xdmxlyo.
17. Caroline Fraser, *Prairie Fires: The American Dreams of Laura Ingalls Wilder* (New York: Metropolitan Books, 2017).
18. Fraser, *Prairie Fires*, p. 507.
19. Ram Trucks, *Farmer*, video, 2:02, posted February 3, 2013, https://www.youtube.com/watch?v=AMpZoTGjbWE. This advertisement ran during the Super Bowl in 2012. According to Wikipedia, Dodge donated to the Future Farmers of America organization "$100,000 for every 1,000,000 views that the YouTube video of the ad received up to $1,000,000. This goal was reached in less than five days."

Chapter 12. Self-Care Is Key

1. Thanks to the Midwest Organic & Sustainable Education Service (MOSES) for making the "In Her Boots" resiliency series available for free.

Chapter 13. Co-farming and Community

1. Chris Smaje, "'Restoration Agriculture' Part II: Annual Monocultures Out-calorie Perennial Polycultures!," Small Farm Future, April 1, 2015, https://smallfarmfuture.org.uk/2015/04/restoration-agriculture-part-ii-annual-monocultures-out-calorie-perennial-polycultures/.
2. There has been a lot of criticism over the past few years of Joel Salatin for racist comments he has made. Those comments are not to be ignored, although they are not a direct part of this narrative. For a full description of the interactions with him and cancellation of his column in *Mother Earth News*, see Tom Philpott's article "Joel Salatin's Unsustainable Myth," *Mother Jones*, November 19, 2020, https://www.motherjones.com/food/2020/11/joel-salatin-chris-newman-farming-rotational-grazing-agriculture/.
3. US Department of Agriculture, Economic Research Service, "Farming and Farm Income," updated February 5, 2021, accessed March 15, 2021, https://www.ers.usda.gov/data-products/ag-and-food-statistics-charting-the-essentials/farming-and-farm-income/.
4. Scott Callahan, "Farmland Ownership and Tenure," US Department of Agriculture, Economic Research Service, November 17, 2020, accessed May 9, 2021, https://www.ers.usda.gov/topics/farm-economy/land-use-land-value-tenure/farmland-ownership-and-tenure/.
5. Wendong Zhang, "Who Owns and Rents Iowa's Farmland?," *Ag Decision Maker*, File C2-78, Iowa State University Extension and Outreach, December 2015, accessed March 18, 2021, https://www.extension.iastate.edu/agdm/wholefarm/html/c2-78.html.
6. Peggy Petrzelka et al., "Understanding and Activating Non-operator Landowners: Non-operator Landowner Survey Multi-State Report," American Farmland Trust, 2020, https://s30428.pcdn.co/wp-content/uploads/sites/2/2020/04/AFT-NOLs-MultiState_Web-5-20b.pdf.
7. Daniel Bigelow and Todd Hubbs, "Land Acquisition and Transfer in U.S. Agriculture," *Amber Waves*, US Department of Agriculture, Eco-

nomic Research Service, August 25, 2016, https://www.ers.usda.gov/amber-waves/2016/august/land-acquisition-and-transfer-in-us-agriculture/.

8. Agrihoods thus far have been mostly limited to very high end housing developments, offering amenities such as golf courses and swimming pools.
9. The article "What Is a Cooperative?" at the website of the University of California's Division of Agriculture and Natural Resources (http://sfp.ucdavis.edu/cooperatives/what_is/) makes no mention, for example, of any Black-led cooperatives in the United States, nor does it list in the early chronology the ways in which Indigenous communities cooperatively owned land and worked farms. The history starts with Benjamin Franklin's Philadelphia Contributionship for the Insurance of Houses from Loss by Fire in 1752 (and lists it as the "oldest continuing cooperative in the U.S.").
10. Livia Gershon, "Yes, Americans Owned Land before Columbus," JSTOR Daily, March 4, 2019, accessed December 21, 2020, https://daily.jstor.org/yes-americans-owned-land-before-columbus/.
11. Kurly Tlapoyawa, "Acequias: A Forgotten History," Mazewalli Nation, Yankwik Mexiko, June 9, 2019, accessed December 23, 2020, https://mesonewmexico.org/2019/06/09/acequias-a-forgotten-history/.
12. Allan Greer, "Commons and Enclosure in the Colonization of North America," *American Historical Review* 117, no. 2 (April 2012): 365–386, https://doi.org/10.1086/ahr.117.2.365.
13. Monica M. White, *Freedom Farmers: Agricultural Resistance and the Black Freedom Movement* (Chapel Hill: University of North Carolina Press, 2018), p. 66, quoting from James C. Cobb, *The Most Southern Place on Earth: The Mississippi Delta and the Roots of Regional Identity* (New York: Oxford University Press, 1992), p. 243.
14. Monica M. White, "'A Pig and a Garden': Fannie Lou Hamer and the Freedom Farms Cooperative," *Food and Foodways* 25, no. 1 (2017): 31, https://doi.org/10.1080/07409710.2017.1270647.
15. White, *Freedom Farmers*.
16. Debbie Elliott, "5 Decades Later, New Communities Land Trust Still Helps Black Farmers," NPR *Morning Edition*, October 3, 2019, https://www.npr.org/2019/10/03/766706906/5-decades-later-communities-land-trust-still-helps-black-farmers.

17. Tadlock Cowan and Jody Feder, "The *Pigford* Cases: USDA Settlement of Discrimination Suits by Black Farmers," Congressional Research Service, May 29, 2013, accessed May 9, 2021, http://nationalaglawcenter.org/wp-content/uploads/assets/crs/RS20430.pdf.
18. The town is listed as 99.44 percent White at http://censusviewer.com/city/IA/Lovilia (accessed May 9, 2021).

Chapter 14. Sharing the Pie

1. The Schwan man traveled the countryside in the 1980s, 1990s, and 2000s selling frozen products. Schwan's Home Delivery service is still available today, although it is far less common.
2. Erin Jordan, "Large-Scale Pork Production May Push Farther into Eastern Iowa," *Cedar Rapids (IA) Gazette*, May 6, 2018, accessed March 14, 2021, https://www.thegazette.com/subject/news/business/large-scale-pork-production-may-push-farther-into-eastern-iowa-20180506.
3. Avinash K. Dixit and Barry J. Nalebuff, *The Art of Strategy: A Game Theorist's Guide to Success in Business and Life* (New York: Norton, 2010).
4. Organic certification was not available in the 1980s. Organic Valley helped craft the laws that are part of certification today.
5. E. Eldon Eversull, "The Long Run: Number of Ag Co-ops Celebrating 100th Anniversaries on the Rise," *Rural Cooperatives* (May/June 2014): 18–23, accessed February 17, 2021, https://localfoodeconomics.com/wp-content/uploads/2018/02/RD_RuralCoopMagMayJun14.pdf.
6. Eversull, "The Long Run," p. 22.
7. Thomas W. Gray and Charles A. Kraenzle, "Problems and Issues Facing Farmer Cooperatives," US Department of Agriculture, Rural Business–Cooperative Service, September 2002, https://www.rd.usda.gov/files/RR192.pdf.
8. Leah Douglas, "How Rural America Got Milked," The Counter, January 18, 2018, accessed January 8, 2021, https://thecounterorg.wpengine.com/how-rural-america-got-milked/.
9. Michael Lee Cook and Molly J. Burress, "A Cooperative Life Cycle Framework," University of Missouri, July 2009, accessed February 12, 2021, https://www.researchgate.net/publication/228545021_A_Cooperative_Life_Cycle_Framework.

10. These numbers were provided by Singh-Watson during our phone conversation.

Chapter 15. People and Policy

1. In 2018, the bill was officially known as the Agriculture Improvement Act of 2018.
2. National Sustainable Agriculture Coalition, "Organic Certification Cost Share," updated May 2019, accessed January 10, 2021, https://sustainableagriculture.net/publications/grassrootsguide/organic-production/organic-certification-cost-share/.
3. Kendyl Landeck, "The Legacy of Redlining and Segregation on Des Moines, Iowa," Iowa State University Digital Repository, Summer 2019, accessed February 23, 2021, https://lib.dr.iastate.edu/creativecomponents/326.
4. Dorothy Schwieder, "History of Iowa," Iowa Official Register, accessed February 26, 2021, http://publications.iowa.gov/135/1/history/7-1.html.
5. Adam Calo, "The Yeoman Myth: A Troubling Foundation of the Beginning Farmer Movement," *Gastronomica* 20, no. 2 (2020): 22, https://doi.org/10.1525/gfc.2020.20.2.12.
6. National Sustainable Agriculture Coalition, "Cultivating the Next Generation: An Evaluation of the Beginning Farmer and Rancher Development Program (2009 to 2015)," October 2017, http://sustainableagriculture.net/wp-content/uploads/2017/10/Cultivating-the-Next-Generation-Oct-2017.pdf.
7. Alan Guebert, "This Should Be Obvious: It's Not 2009," *Columbia (MO) Daily Tribune*, January 8, 2021, accessed February 28, 2021, https://www.columbiatribune.com/story/opinion/columns/2021/01/08/farm-and-food-file-should-obvious-its-not-2009/6570166002/.
8. Shoshana Inwood et al., "Health Insurance and National Farm Policy," *Choices* 33, no. 1 (1st Quarter 2018): 5, https://www.choicesmagazine.org/UserFiles/file/cmsarticle_614.pdf.
9. Alana Knudson told me this during a phone interview on November 6, 2020.
10. US Department of Agriculture, "2017 Census of Agriculture Data Now Available," press release no. 0051.19, April 11, 2019, accessed February 28, 2021, https://www.usda.gov/media/press-releases/2019/04/11/2017-census-agriculture-data-now-available.

11. Sonny Perdue, "Report to the President of the United States from the Task Force on Agriculture and Rural Prosperity," US Department of Agriculture, October 21, 2017, accessed February 28, 2021, https://www.usda.gov/sites/default/files/documents/rural-prosperity-report.pdf.
12. Federal Communications Commission, "Task Force for Reviewing the Connectivity and Technology Needs of Precision Agriculture in the United States," accessed May 10, 2021, https://www.fcc.gov/task-force-reviewing-connectivity-and-technology-needs-precision-agriculture-united-states.
13. C Spire Rural Broadband Consortium, "Understanding the Rural Broadband Problem," 2018, p. 7, accessed February 28, 2018, https://www.cspire.com/resources/docs/rural/ruralbroadband-whitepaper.pdf.
14. Richard Florida and Adam Ozimek, "How Remote Work Is Reshaping America's Urban Geography," *Wall Street Journal*, March 5, 2021, accessed March 15, 2021, https://www.wsj.com/articles/how-rmote-work-is-reshaping-americas-urban-geography-11614960100?tesla=y.

Chapter 16. Meanwhile, Back at the Ranch

1. Charles D. Reed, "Iowa Droughts," *Bulletin of the American Meteorological Society* 15, no. 10 (October 1934): 215–218, accessed March 2, 2021, http://www.jstor.org/stable/26263211.
2. Justin Gehrts, "Record Heat Hit Iowa on This Date in 1934," KCRG-TV9, August 8, 2016, accessed March 8, 2021, https://www.kcrg.com/content/news/WWC-Record-heat-hit-Iowa-on-this-date-in-1934-389505271.html.

Acknowledgments

First and foremost, I would like to thank Leroy Hogeland, who gave so much of his time to tell me stories about his life. Leroy, you entrusted me with your family stories, and I hope I did them justice.

Thanks too to Dorothy Lynn Hogeland, who also shared important perspectives on these topics and taught me a lot. I wish you were still here to read this book and to see all we have done on the farm. You would be proud.

To John's sister Andrea, her husband, Jeff, and son, Lucas, thanks for putting us up in your basement and for all the support over the years. Also thanks to Roma, who listened to many of Leroy's stories as we sat at the kitchen table, and to my brother, Doug, who helped make our new house a home. To the boys, you two taught me that Iowa and the farm are wonderful places. I appreciate everyone in the Hogeland and Sherman clans, who have given us so much love and support in our farm journey too.

I would like to thank everyone who gave of their time to be interviewed and those whose research and writing taught me so much. Thank you to all those who came before us in building cooperatives, nonprofits, farmer groups, granges, and unions, those who tried interesting and often successful models on which we can grow and improve our world.

Thanks to Felicia Olivera for all the editing work, and to Emily Turner as well.

A very special thanks to my research assistant and biggest fan, Enid Hoffman, aka, Mom.

And last and of course not least, to my proofreader, chef, and all-around best husband ever, John. Tons of thanks and love.

About the Author

Beth Hoffman is a beginning farmer on 530 acres in Iowa. For the past twenty years, she has worked as a journalist covering food and agriculture. Her work has been aired and published on NPR's *Morning Edition*, the *Guardian*, *The Salt*, *Latino USA*, and the *PBS NewsHour*.

Island Press | Board of Directors